HISTORY *of* THE FIRE COMPANIES *of* FREDERICK COUNTY MARYLAND

HISTORY *of* THE FIRE COMPANIES *of* FREDERICK COUNTY MARYLAND

CLARENCE "CHIP" JEWELL

THE History PRESS

Published by The History Press
Charleston, SC
www.historypress.com

Front cover image: Sherry E. Kemp.

First published 2021

Manufactured in the United States

ISBN 9781467149761

Library of Congress Control Number: 2021937194

Notice: The information in this book is true and complete to the best of our knowledge. It is offered without guarantee on the part of the author or The History Press. The author and The History Press disclaim all liability in connection with the use of this book.

Dedicated in memory of Lydia Butterworth, John Maly and Jim "Sonny" Ridgley, who were members of the Frederick County Fire and Rescue Museum and Preservation Society that helped save and preserve the history of the Fire and Rescue Service of Frederick County, Maryland.

This book is also dedicated to the artist of the magnificent painting on the cover of the book. Sherry Kemp was a talented artist who allowed her work to be used on the cover of this book and passed away just prior to the book going to print.

Special thanks to

Jim and Joy Deater, Jim Graham, Angie Hummer, Chief George Clary, Jim Dixon, Mayor John Kinniard, Vaughn Zimmerman, Chris Beard, Donald Trimmer, Dave Snyder, Jeep Abrecht, Steve Hurst, Jack Fleishman, Sherry Kemp, Chief Tom Owens, Pastor Emeritus Fred Wenner, Frederick County Fire and Rescue Museum, Heritage Frederick (formerly the Historical Society of Frederick County), all the volunteer fire and rescue companies in Frederick County that shared their individual histories and especially my wife, Kathy, who supported me in this endeavor.

All proceeds benefit
Frederick County Fire and Rescue Museum and Preservation Society, Inc.
300B South Seton Avenue
Emmitsburg, MD 21727

CONTENTS

Contents

FOREWORD

In February 2010, I was appointed to serve as the fire chief/director of fire and rescue services in Frederick County, Maryland. One of the very first people to introduce themselves to me was then director of Frederick County Emergency Communications Chief Clarence "Chip" Jewell. The first thing he did when we met was place in my hand copies of the two previous Frederick City/County fire service Arcadia Publishing *Images of America* history books he had written or co-authored: *Firefighting in Frederick* (2004) and *Firefighting in Frederick County* (2005). These two books were of tremendous value to me because they introduced me to the rich history of the establishment and growth of the fire and emergency medical services in both Frederick County and the City of Frederick. Chief Jewell's first two books, as well as this new addition to the collection, also made me realize how well the Frederick fire service cares for and preserves the history of both the individual fire companies and the greater fire service of the city and county combined.

As time passed, Chip and I learned that our own personal histories in the fire service had many parallels, particularly our volunteer fire service experiences. Both of us are lifelong volunteer firefighters with more than fifty years of service. We had both grown up in very active volunteer stations and both entered the career fire service by starting in emergency communications, before the days of 911 and Computer Aided Dispatch. We would talk of hair-raising experiences of "riding the back step," the many challenges of performing CPR in the back of a Miller-Meteor

Cadillac Ambulance or the days of being the only dispatcher on duty and having to answer the emergency phone located in the bathroom while managing our bodily functions. These shared experiences made for some interesting lunch table conversations as we both recounted similar stories from our past. The more Chip and I talked, the greater the appreciation I had for his love of history in general and fire service history in particular. I found that as a son of Frederick, Chip has invested a great deal of himself to study the global history of Frederick City, from the days of its earliest German settlers. He often writes contributing articles on Frederick's rich history for the local newspaper.

In 2014, Chip became part of Frederick fire history himself when he was appointed to be the first director/chief of Volunteer Fire-Rescue Services for the newly established Frederick County Division of Volunteer Fire-Rescue Services. In this role, Chief Jewell was responsible for the oversight and coordination of the twenty-five volunteer fire-rescue companies within the fire-rescue system. Within the operational chain of command, Chip was a uniformed deputy fire chief and shift commander. In this position, Chief Jewell and I worked very closely together on a day-to-day basis, and this placed me in a position of being able to learn even more about the history of not only the Frederick City and County services but also of many of the specific volunteer fire company histories that Chip had studied and worked to help preserve.

In this book, the depth of his research and knowledge is impressive because he explains the many tentacles of the greater Maryland fire service from the days of the Ancient and Honorable Mechanical Company of Baltimore and how they are connected to the Frederick fire service. He provides insight into the challenges the early companies had in raising the funds needed to purchase apparatus and equipment and the creative ways this was accomplished. It seems some things never change.

Chief Jewell further relates the many ways the early volunteer fire companies of Frederick—Independent, Junior and United—served both their communities and their country. It is fascinating to learn how and why it was the volunteer fire companies that formed the first local militia units to provide security for their communities and even more fascinating to learn how the Frederick volunteer firefighter militia units were instrumental in quelling the insurrection by abolitionist John Brown during his raid on Harpers Ferry, Virginia (now West Virginia). The book expands on the role these firefighter militia units played during the course of the Civil War, especially in a town that had a split allegiance between North and South.

Foreword

This book will take you on a journey that reveals how the great rivalries of the early Frederick fire companies evolved into a more cooperative relationship among the companies and how Frederick was the catalyst for the formation of the Maryland State Firemen's Association.

Chief Jewell expands on Frederick's fire service history as he explains how growth in Frederick led to the formation of the Citizens Truck Company and how the towns outside Frederick led to the formation of additional fire and ambulance companies over the years. He shares the details of how companies in the communities of Middletown, Jefferson, Braddock Heights, Myersville, New Midway, Rocky Ridge, Adamstown, New Market, Woodsboro, Graceham, Brunswick, Wolfsville, Lewistown, Urbana, Walkersville, Thurmont and Emmitsburg were formed. He details their founding leaders and how they acquired their initial apparatus and equipment.

For those who love stories of significant fires that occurred in the early days, Chief Jewell does not disappoint. His research for this book reveals fascinating details of major fires that occurred both in Frederick proper and in surrounding communities. In many instances, a fire in a smaller community would require help from Frederick to be summoned. It is fascinating to read the details of how Frederick's steam engines were loaded onto train cars for transportation to an out-of-town fire.

According to the dictionary, a historian is "an expert in or student of history, especially that of a particular period, geographical region, or social phenomenon." There is no question that Chief Jewell fits this definition. Chip is a lifelong student of history, and his passion for history resides in its many forms:

He is a researcher who strives to ensure historical accuracy in his writings and has now authored his third book on the history of the Frederick fire services and many articles on the history of Frederick, Maryland, in general.

He has and continues to invest himself personally in the history of the fire service through his involvement in the Frederick County Fire Rescue Museum, located in Emmitsburg, Maryland. Chip is a founding member of the museum, was a driving force for its establishment and is currently president of the museum dedicated to "protect, preserve and display the artifacts of the history and tradition of the fire and rescue service of Frederick County, Maryland."

He is a student of the living history of the Frederick County Fire and Rescue Service. For every career recruit training academy class, every volunteer fire-rescue new member orientation class, for civic organizations and others, Chief Jewell has developed a detailed living history presentation where, clad in a historical uniform of the Independent Hose Company, he delivers an oratory of the history of the formation of the fire service in Frederick that captures the attention of all who attend his presentation.

It is an honor for me to write the foreword for my friend and colleague's latest work. Anyone who has an interest in history in general or in Frederick County, Maryland, specifically will find this book to be a fascinating look into the early history of this community's fire service.

—Chief Thomas W. Owens, CFO
Fire Chief/Director of Fire-Rescue Services (Retired)
Frederick County, Maryland

Chapter 1
EARLY FIRE PROTECTION IN THE CITY OF FREDERICK

The early German settlers in Frederick County who came to America from the Rhine River Valley followed the Monocacy Trail and discovered an area very similar to their former home of Schifferstadt, Germany.[1] The Catoctin Mountains, Monocacy River Valley and relatively mild climate made Frederick the ideal area for German settlement. Fredericktowne, later Frederick, was established in 1745. Only fifteen years later, the town fathers determined the need to purchase a fire engine. In 1760, a raffle was held to raise money to build a marketplace and to purchase the first fire engine.[2] The engine was purchased in 1764 and was named "Grandfather." It is believed this very first engine was used by the Frederick Hose Company and later by the Independent Hose Company No. 1.[3]

The purchase of this first fire engine in Frederick makes Frederick the third-oldest fire service in Maryland, after Chestertown and Baltimore. The Ancient and Honorable Mechanical Company of Baltimore was organized in 1763 and for several years also served as police and magistrate, as well as firemen.[4] A much deeper relationship between the Mechanical and Frederick will be highlighted in a later chapter. On December 26, 1794, the Corporation of Frederick passed an ordinance that required all homes to keep leather buckets to supply water in case of fire.[5] A leather fire bucket that was owned by the Schley family, early settlers in Frederick, is now owned by the Junior Fire Company No. 2 and is displayed at the fire station.

Evidence of several early fire companies provides a brief glimpse of the early days of Fredericktowne and the efforts at fire protection. In 1817,

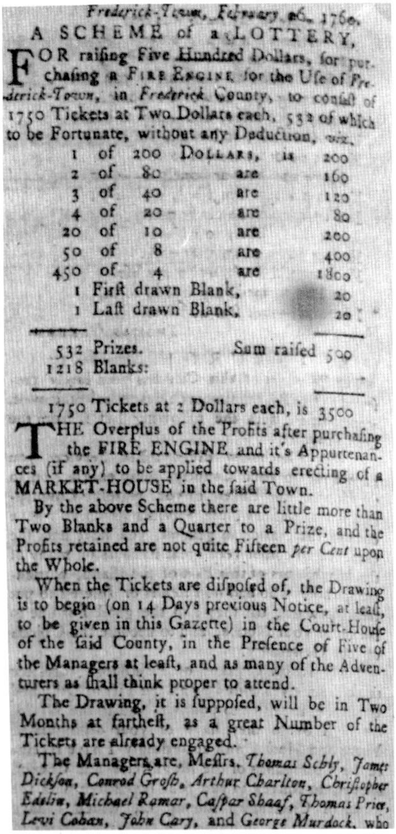

Frederick-Town, February 26, 1760.

A SCHEME of a LOTTERY,

FOR raising Five Hundred Dollars, for purchasing a FIRE ENGINE, for the Use of Frederick-Town, in Frederick County, to consist of 1750 Tickets at Two Dollars each, 532 of which to be Fortunate, without any Deduction, viz.

1	of	200	Dollars, is	200
2	of	80	are	160
3	of	40	are	120
4	of	20	are	80
20	of	10	are	200
50	of	8	are	400
450	of	4	are	1800
1	First drawn Blank,			20
1	Last drawn Blank,			20
532	Prizes.		Sum raised	500
1218	Blanks:			
1750	Tickets at 2 Dollars each, is			3500

THE Overplus of the Profits after purchasing the FIRE ENGINE and it's Appurtenances (if any) to be applied towards erecting of a MARKET-HOUSE in the said Town.

By the above Scheme there are little more than Two Blanks and a Quarter to a Prize, and the Profits retained are not quite Fifteen *per Cent* upon the Whole.

When the Tickets are disposed of, the Drawing is to begin (on 14 Days previous Notice, at least, to be given in this Gazette) in the Court-House of the said County, in the Presence of Five of the Managers at least, and as many of the Adventurers as shall think proper to attend.

The Drawing, it is supposed, will be in Two Months at farthest, as a great Number of the Tickets are already engaged.

The Managers are, Messrs. *Thomas Schly*, *James Dickson*, *Conrod Grosh*, *Arthur Charlton*, *Christopher Edelin*, *Michael Ramar*, *Caspar Shaaf*, *Thomas Price*, *Levi Cohan*, *John Cary*, and *George Murdock*, who

Left: A copy of the 1760 lottery announcement to raise money to purchase the first fire engine in Frederick. *Courtesy Frederick County Fire and Rescue Museum.*

Below: A fire bucket owned by the Schley family, one of the first families in Frederick. It is now owned by and displayed at the Junior Fire Company. *Author's photo.*

Double yoke fire buckets used by Independent Hose Company No. 1. *Courtesy Independent Hose Company No. 1.*

an ordinance was passed that established fire wards. The First Ward was the Friendship Fire Company, Second Ward the Union Fire Company, Third Ward the Frederick Hose Company and Fourth Ward the Sun Fire Company. All white taxable males were declared to be members of the fire company located in the ward in which they lived.[6] The fire wards were separate and distinct from the early city election wards, with each of the four fire companies assigned specific areas of the city.[7]

There is minimal documentation so far discovered of the early fire companies in Frederick. Though the history of the Independent Hose Company indicates the Frederick Hose Company was a predecessor, a notice in the *Frederick Herald* on November 5, 1831, advertises a meeting of the First Frederick Hose Company at Mrs. Michaels Tavern at six o'clock—fourteen years after the Independent Hose Company was organized. The notice was authority of William H. Albuagh, secretary. Another meeting notice was also posted in the *Frederick Herald* on March 3, 1832.[8]

The January 4, 1909 *Frederick News* had an article on the discovery of papers found in the courthouse files for the Sun Fire Company to conduct a lottery in the amount of $875 to purchase a fire engine.[9] Though the article notes there had been no references to the Sun Fire Company prior to this document finding, the Archives of Maryland, Volume 0607, Session

Laws of 1805, provide the actual legislation passed by the Maryland state legislature approving the lottery. The law also states Henry Steiner, Francis Mantz, Henry Bantz, George Creager Sr., Lewis Wilzshimer, Charles Schell, Jacob Medtard, George Adam Ebert and Peter Burkhart must execute a bond to the State of Maryland in the amount of $1,750 as a condition to conduct the lottery.[10] In response to the January 4 article in the *News*, Dr. Henry Petit, a resident of Frederick from 1847 to 1854, said the Sun Fire Company was located "at the east end of the Market House almost against the wall of the old Lutheran graveyard....In my younger days [there] stood an old red brick building used as a tool house. In this was kept the Sun fire engine. It had brakes like an ordinary engine, was small about the size of an old fashioned six octave piano."[11]

The only information presently known of the Friendship Fire Company of Ward 1 was that the new Young Men's Fire Company, ultimately organized as the Junior Fire Company, took possession of the Friendship Fire Engine and petitioned individuals, insurance companies and the city council for funds.[12] An article in the *Frederick Examiner* speaks of a brawl that occurred on August 14, 1838, between three fire companies, one of which was reported as the Friendship Fire Company. No other documentation of the Friendship being in existence at such a late date has thus far been discovered.[13]

The Independent Hose Company No. 1 is the oldest continually operating fire company in the state of Maryland. The company has provided continuous service since inception in 1818; however, it was not incorporated until February 27, 1840.[14] A cornerstone taken from the original engine house bears the date of 1816, which predates the present-day Independent Hose Company No. 1 and may have been the location of an earlier fire company. In 1827, the first Independent Hose Company engine house was built on the Court House Square in the location of the old hay scales. The building was built by John Shipman for a cost of $550, paid by a Levy Court of Frederick County.[15]

On August 19, 1845, the Independent Hose Company purchased land from Lawrence J. Brengle and his wife on West Church Street next to the German Reformed Church, today known as the Trinity Chapel or "Town Clock." The land was purchased for a cost of $750 and provided the fire company a street frontage of thirty-one feet. Ground was broken on August 3, 1846, for a two-story fire engine house and meeting room.[16] For most of the over two-hundred-year history of the Independent Hose Company, the fire station was located along West Church Street until moving to Baughman's Lane in 1978.

In May 1847, the Independent Hose Company purchased a bell from Philadelphia for a cost of $80. On December 4, 1849, a new larger bell was installed with the inscriptions of "E. Shriver President—G.J. Fischer Vice J. McSherry Treas. H.C. Steiner Secy." The bell was made by Clampitt & Register Baltimore and weighed 563 pounds. The bell inscription also included "Independent Hose Company. Formed 1818—Incorporated 1838." This bell is still owned by the Independent Hose Company No. 1 in Frederick.[17]

The earliest apparatus used by the Independents was a "small hand engine," most probably the Grandfather. This engine was also believed to originally be used by the Frederick Hose Company. On March 9, 1827, Jacob Engelbrecht noted in his diary that the Independent Hose Company received a new fire engine. "It was made by Pressrs. Merrick and Agnew Philadelphia & cost eight hundred dollars. They have also five hundred feet of copper riveted hose."[18] The diary also noted the new engine was moved into the new firehouse on September 22, 1827, the day the first company meeting was held in its new quarters. On October 30, 1841 the Independent Hose Company received a new "suction" engine built by John Agnew of Philadelphia for a cost of $900.[19]

In 1860, the Independents purchased an 1843 "gallery" engine from the City of Baltimore for $450. This engine was built by the John Rodgers Company of Baltimore and arrived in Frederick on September 15, 1860. The engine was originally owned by the Deptford Fire Company located Baltimore until the volunteer fire companies in the city of Baltimore were disbanded and the Baltimore City Fire Department was established. This engine was one of three owned by the Deptford volunteer company. The Deptford Fire Company No. 4 was organized in 1792 and located on Broadway Fells Point.[20] On September 15, 1860, the engine was tested by throwing a stream of water 184 feet.[21] On April 13, 1860, "Mr. T. Edgar Ramsburg was run over by the Independent Hose Reel." The Independents were responding to what was determined a false alarm at the corner of Market Street and Patrick Street. Ramsburg was "shaken & not as much injured as was supposed last night."[22]

In 1836, an act was passed in the Maryland legislature that exempted those seventy years and older from attending fire company meetings and assisting at fires. It is noted this act was passed prior to the Junior Fire Company being organized, thus indicating additional fire companies were still in existence in 1836.[23]

> *An act for the relief of the old citizens of Frederick town in Frederick county. Passed Jan. 25, 1836*
>
> *Be it enacted by the General Assembly of Maryland, Relief extended that all and every citizen or resident of, or in Frederick city in Frederick county, who are, or shall be of the age of seventy years, shall be, and they are exempt from all and every duty in relation to fire, company meetings and fires; and shall not be required by the corporate authorities of Frederick city, or the fire companies thereof, to attend any such meetings, or to be present and assisting at fires, and shall not be fined for any neglect to attend said meetings or fires.*

In November 1838, a group of young men gathered at Mantz's Drug Store in Frederick to discuss organizing another fire company to protect Frederick. Afterward, a notice in the *Herald* invited "all young men of this town that would be interested in uniting with an additional fire company for the town of Frederick." The company also became known locally as the Young Men's Fire Company as young men of the community were called on to organize the company. Junior Fire Company No. 2, as known today, was founded on December 6, 1838. The company was officially incorporated as the Junior Fire Company of Frederick in the legislative session that ran from December 30, 1839, until March 21, 1840.[24]

In January 1839, the fire company ordered a hand pumper from the John Rodgers Company of Baltimore for the cost of $1,100. The company had a choice of two models of the Rodgers pumper, the Junior or the Senior. The Junior model was selected, which made the fire company name Junior even more appropriate. The company also purchased leather hose from Dukeharts of Baltimore and hose reel from Henry Boteler of Frederick.[25] Recently, Jim and Joy Deater, authors of the 175th anniversary book of the Junior Fire Company, discovered an article from the *Gettysburg Complier* dated April 3, 1879, that mentioned the Junior engine was sold to Littlestown, Pennsylvania, when a fire company was organized in that Pennsylvania community.[26]

The engine arrived in Frederick on Saturday, August 24, 1839. Engelbrecht's diary notes, "Several companies of this town escorted them [the Juniors] from the railroad depot through the principle [*sic*] streets." The reference to "several companies" indicates that at least some of the fire companies noted earlier in this chapter were still in existence in 1839. In addition, the Washington Hose Company, organized in 1837 and discussed

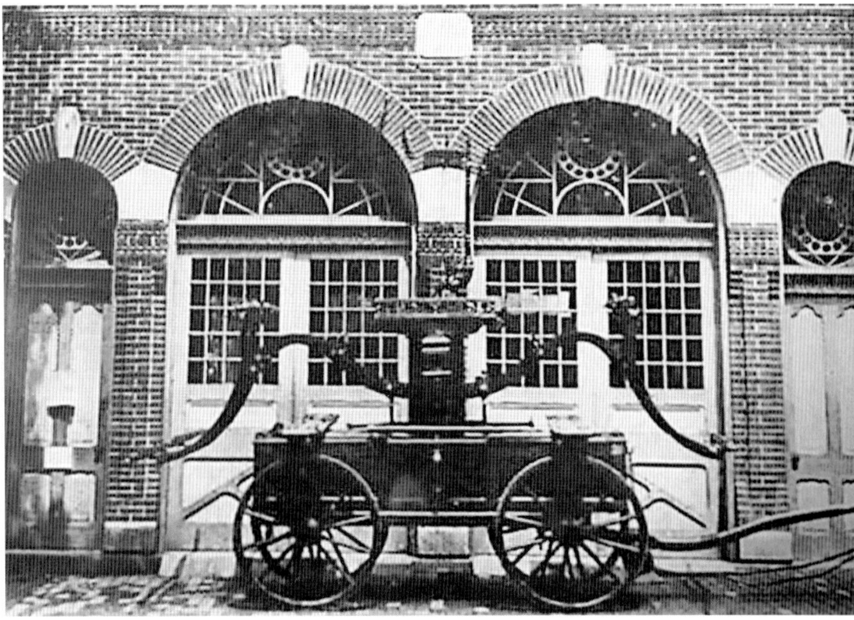

The 1839 John Rodgers original engine of Junior Fire Company No. 2 after being sold to Littlestown, Pennsylvania. *Courtesy Jim and Joy Deater.*

further in this chapter, would have also been in existence. On Monday, August 26, 1839, the new engine was tested when it threw water over the steeple of the Evangelical Reformed Church steeple "a little higher than the extreme knob, which was at least 150 feet."[27] This test over the Town Clock was probably the first of many fire companies testing engines by throwing water over the Evangelical Reformed Church Steeple.[28] This engine is believed to have been sold to Littlestown, Pennsylvania, and later donated to the Smithsonian Institute.

In the Archives of Maryland, Session Laws, 1842, an act was passed authorizing the sum of $500 to be appropriated to the presidents of the Independent Hose Company and the Junior Fire Company of Frederick "to defray expense incurred in repairing their engine or engines, and to aid said companies in purchasing such apparatus, as may be needful for their better and more effectual equipment."[29] There is no indication that either company utilized the funds to purchase new apparatus.

The year 1840 marks the first officially recognized line-of-duty death of a firefighter in the state of Maryland. William Carlton, a city registrar and son of Thomas Carlton, an early mayor of Frederick, collapsed while pulling the Junior engine to respond to a fire at the residence of Colonel

John McPhearson on Court Street. Engelbrecht's diary reported, "He became exhausted, fell down and in a short time thereafter expired."[30] He was only in his early twenties and had been married only about six months to Susan Niell of Philadelphia.[31] Carlton was a founding member of the Junior Fire Company and is the first name enshrined on the Maryland Fire-Rescue Memorial in Annapolis, Maryland, and the first line-of-duty death listed on the Hall of Honor at the Frederick County Fire and Rescue Museum in Emmitsburg.

The first engine house of the Junior Fire Company was located at the intersection of Market and Second Streets on land donated by the Farmers and Mechanics National Bank. The lot was located behind the bank and faced Second Street with twelve feet of Second Street frontage and a depth of eighty-five feet.[32] The second fire hall was built on a lot purchased from the heirs of Peter Hardt at a public sale for $1,825. The building housed other businesses and organizations. The building also included a large social hall with a stage. The large hall was used as part of General Hospital #6 during the Civil War.[33]

On June 6, 1851, the Junior Fire Company completed the cupola on the engine house, also known as the Temperance Hall as the Sons of Temperance was one of the occupants. Weighing 723 pounds, a bell to alert volunteers

Memorial of William Carlton at the Frederick County Fire and Rescue Museum in Emmitsburg. *Author's photo.*

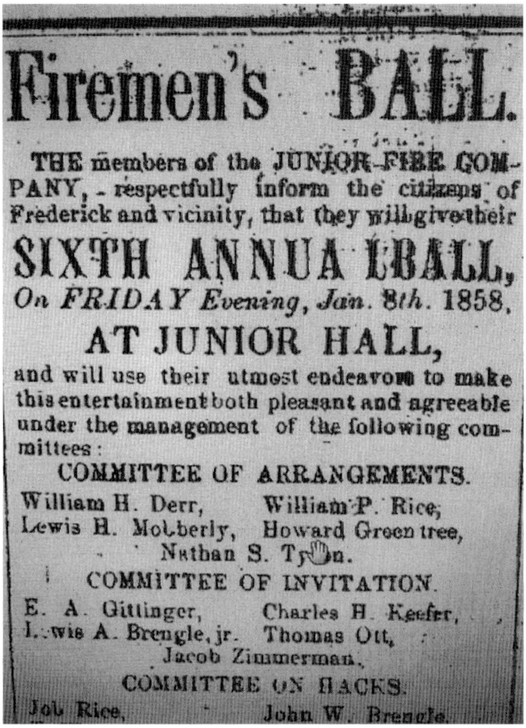

Firemen's BALL.

THE members of the JUNIOR FIRE COMPANY, respectfully inform the citizens of Frederick and vicinity, that they will give their

SIXTH ANNUAL BALL,

On FRIDAY Evening, Jan. 8th. 1858,

AT JUNIOR HALL,

and will use their utmost endeavors to make this entertainment both pleasant and agreeable under the management of the following committees:

COMMITTEE OF ARRANGEMENTS.

William H. Derr, William P. Rice,
Lewis H. Moberly, Howard Greentree,
Nathan S. T[illegible]n.

COMMITTEE OF INVITATION.

E. A. Gittinger, Charles H. Keefer,
Lewis A. Brengle, jr. Thomas Ott,
Jacob Zimmerman.

COMMITTEE ON HACKS.

Job Rice, John W. Brengle.

Firemen's Ball at the Junior Hall advertisement in the *Frederick Examiner*, January 8, 1858. *Courtesy Jim and Joy Deater.*

was made by A. Meneely & Sons of West Troy, New York, and installed in the cupola. On May 30, 1851, the fire company installed the adornment *Goddess of Liberty* on the top of the cupola.[34]

In 1852, the Junior Fire Company altered the building by installing winding stairs at the front door, strengthening the joist work and adding a theatrical stage to the east end of the room. To add to the ambiance, the room was "frescoed" by Earnest Dryer of Baltimore. A ball was held on April 21, 1852, to help defray the cost of the alterations. Tickets sold for two dollars each. The building was expanded in 1872.[35]

In 1837, the Washington Hose Fire Company was organized by the citizens of Fire Ward No. 2. In 1817, Fire Ward No. 2 was initially assigned to the Union Fire Company. In 1837, an ordinance passed by the aldermen and Common Council of Frederick stated all taxable white male citizens residing within the limits of Ward No. 2 were compelled to join the fire company of the ward. The ordinance also levied fines for not attending fire company meetings in the amount of twelve and a half cents—"no excuses." If the member failed to pay the fine, a bill would be sent in the name of the "Mayor, Aldermen and Common Council of Frederick."[36]

A parade belt of the Washington Hose Company on display at United Steam Fire Engine Company No. 3. *Author's photo.*

Members of the company that participated in the bucket brigades were called "lane men." Other members were assigned as "property guards" and "bag men." The bag men carried large linen bags that property could be put in to remove from a fire. There is no record of the Washington Hose Company owning a pump engine; it is noted that the company owned a "water tank, at which buckets were filled with water during a fire." The metal water tank that is tandem with the United hand pumper "Old Lady" on display at the Frederick County Fire and Rescue Museum in Emmitsburg most likely is this original water tank of the Washington Hose Company.[37]

When first organized, the Washington Hose Company was headquartered on East All Saints Street.[38] Later Sanborn Insurance maps indicated an "enginehouse" was located on South Bentz Street near present-day Gaither Alley that may have been another location of Washington Hose or another unnamed Frederick fire company. The Washington Hose Company disbanded in 1844, after pleading with the Independent Hose Company to take the company on as an auxiliary. At a meeting of the membership of the Independents on March 3, 1844, the Independents voted not to accept the Washington Hose Company as an auxiliary company. As a result, the Washington Hose Company ceased to exist.[39] The area around All Saints Street and Market Street would not be without fire protection very long.

On November 2, 1845, some former members of the Washington Hose Company met at the firehouse with the intention of forming a new fire company. It was agreed to name the new company the Mechanics Hose

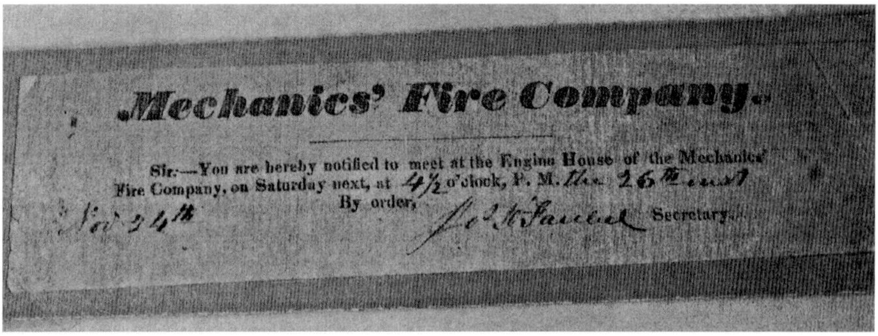

Mechanics' Fire Company.

Sir:—You are hereby notified to meet at the Engine House of the Mechanics' Fire Company, on Saturday next, at 4½ o'clock, P. M. the 26th inst.

By order,

Secretary.

Nov 24th

Top: A metal "Tank" believed owned by the Washington Hose Company and used by United Steam Fire Engine Company No. 3 presently on display at Frederick County Fire and Rescue Museum. *Author's photo.*

Bottom: An announcement of a meeting of Mechanics' Fire Company on November 26, 1845. *Courtesy United Fire Company archives, photo by author.*

Company. However, at the next meeting held on November 26, 1845, the name was changed to the United Hose Company. With ninety-two members signing the original constitution, the new company formally organized and met every Saturday. The United Company took over the former Washington Hose Company building and tank.[40]

In 1847, the City of Frederick provided $430 toward the purchase of a lot on what today is 79 South Market Street, United Steam Engine Fire

The original cupola and "Swamp Bell" as they appear today with mascot weather vane "Johnny Swamper." *Author's photo.*

Company No. 3. The ground was broken for a new building in 1848. When completed, the building was sixty feet deep. The original section of the building, bell and cupola still exist. The building is believed to be the oldest continuously operating fire station in America, having provided continuous response to fires since first occupied in 1848.[41]

The firehouse was built in a very swampy area near Carroll Creek; thus, the firehouse was dubbed "Swamp Hall" and the firefighters of the Uniteds got the nickname "Swampers." The colorful figure atop the cupola is the company mascot "Johnny Swamper." Even the bell atop the cupola was locally known as the "Swamp Bell," as noted in Engelbrecht's diary.[42] The bell was also sounded to warn the citizens of flooding of Carroll Creek. Initially, the word "Swampers" adorned the company hats, but a vote in 1850 changed the logo to "1845," the year the company was organized.[43]

One of the first fires that the new company responded to was on the early morning of July 26, 1846, at the stable of Dr. William Bradley Tyler. "The Swamp Rangers were there first with their hose, which put it out." Engelbrecht reported about three-quarters of the roof and some upper loft was burned with an estimated loss of about $150.[44] An entry in Engelbrecht's diary on February 1, 1851, noted, "The United Hose (Swampers) got a new

A parade hat, circa 1850, after United Fire Company voted to replace "Swampers" with the organization year 1845. *Courtesy of United Fire Company archives, photo by author.*

engine this week paraded through the streets for its power.—The company was escorted through the town by Junior and Independent Fire Companies—The engine was built in Philadelphia by Merrick & Agnew and cost $1100."[45] A new hose reel was purchased in 1854 for a cost of $125.[46]

In response to a public hanging that took place on Cannon Hill in 1858, members of the United Fire Company organized the United Guards, a militia unit that used the fire hall for meetings and storage of armaments. On March 19, 1858, forty men paraded through the streets of Frederick. A week later, the unit was inspected by Colonel Shriver, commander of the Maryland Regiment of the Maryland National Guard, the military to which the United Guard were attached.[47] Also in 1858, the American Band, also associated with the United Fire Company, was organized. The band often accompanies the fire company to parades and celebrations.[48]

A notation in the Engelbrecht diary also indicates a "hook and ladder" company was organized in August 1879. The president of the company was Frank B. Carlin, and it was noted that the company "had not received the machine yet that year."[49]

Chapter 2

THE MYSTERY OF THE "OLD LADY"

The return of the "Old Lady" back to Frederick County is a story unto itself. The original Old Lady was an engine built in 1821 for the Mechanical Company No. 1 of Baltimore City by Philadelphia fire engine builder Perkins and Bacon for the cost of $1,050.[50] As background, the Mechanical Company No. 1 of Baltimore was originally organized in 1763 and formally known as the Ancient and Honorable Mechanical Company of Baltimore. The Mechanical Company essentially discharged the duties of the town such as the first school, first sheriff, first hospital and first fire company. This organization still exists today and lays claim as the oldest civic organization in the United States.[51] Ultimately, the City of Baltimore was incorporated and assumed the normal duties of a municipality except fire protection, which the Mechanical Company retained, becoming Mechanical Company No. 1 and purchasing the first fire engine in 1769.[52]

The Uniteds received the mahogany engine Old Lady from the Mechanical on April 19, 1860, for a cost of $400.[53] As this engine would have been owned by the United Fire Company in 1861, it is highly probable the Old Lady fought the May 8, 1861 courthouse fire the weekend after the Maryland legislature met in Frederick to defeat an attempt at succession from the Union.[54]

However, the story of the Old Lady continues for another 150 years. After the Uniteds purchased the "Lily of the Swamp," an 1878 Clapp and Jones Steam Engine discussed in greater detail later, the Old Lady found its way to Libertytown. According a transcript copy of the April 1, 1957 meeting

The original gallery Old Lady engine purchased in 1860 from the Mechanical Fire Company No. 1 of Baltimore. *Courtesy of United Fire Company archives.*

of the United Fire Company, the Old Lady was loaned to Libertytown to fight fires in that community.[55] In 1899, a fire company in Libertytown was formed with the use of the Old Lady as the town fire engine. After a major fire in Libertytown, the May 20, 1899 edition of the *Daily News* reported that a meeting of the citizens of Liberty was held to organize a fire company and have the hand engine repaired. On August 18, 1899, the paper reported the old engine was ready for service and the fire company organized.[56]

The Old Lady remained in Libertytown until February 9, 1932, when the pumper and water tank were sold by the Libertytown Civic Improvement Association to George Conner for a display of the museum at the Baltimore Equitable Insurance Company. According to the article in the *News*, the pumper and tank were purchased from the United Fire Company in 1868. The article also noted the town built a dam device along the town creek to provide water for firefighting.[57] The information about Libertytown purchasing the Old Lady is in conflict with the records that state the engine was "loaned." It is unlikely the United Fire Company would sell the engine before 1878, as the Old Lady was most probably the largest pumper owned by the Uniteds. Thus, the Old Lady probably would not have been sold until the Lily of the Swamp was in service. Possibly, the correct date, if the engine

Top: The suction engine called Old Lady at the United Fire Company of Libertytown quarters, circa 1900. *Author's collection.*

Bottom: The suction engine Old Lady as it appears today at the Frederick County Fire and Rescue Museum *Author's photo.*

was sold, would have been after 1878, the year the Lily was purchased. Regardless, after being sold to Conner, the Old Lady and metal tank were ultimately displayed at a museum of the Equitable Insurance Company in Baltimore until sold at an auction to the Granite Hand Tub Association of Newmarket, New Hampshire. The engine finally returned to Frederick

The suction engine Old Lady throwing water on the former Frederick County Courthouse during a reenactment of the 1861 courthouse fire in May 2011. *Courtesy Frederick County Fire and Rescue Museum.*

on October 29, 2006. The Old Lady as described is now on display at the Frederick County Fire and Rescue Museum in Emmitsburg.

The work of several museum members, especially Frank Schmersal, Doc Kinney and Sonny Ridgley, was completed in time for the 150th anniversary of the 1861 courthouse fire. A community bucket brigade was formed, and the Old Lady threw water on the same building she most likely threw water on to extinguish a fire almost exactly 150 years prior.[58]

The Old Lady is a puzzle unto itself. Many unanswered questions are still shrouded in mystery. The arrival of the Old Lady is documented, but the question remains as to what the Old Lady truly is. The original Old Lady of the Mechanical Company No. 1 was dismantled in 1845 and rebuilt by the John Rodgers Company of Baltimore. Wooden parts of the original engine were made into wooden canes for the older Mechanical members. This rebuilt Old Lady was also an ornate gallery engine.[59] The minutes of the United Fire Company membership meeting on July 1, 1886, state, "Moved that the Gallery engine which formerly belonged to the old Mechanical

Engine be given to M. Dukehart. Motion laid over to the next regular meeting." Further notes from the October 4, 1886 meeting state, "Motion was amended to substitute the Veteran Firemen's Association of Baltimore and passed." Thus, this information indicates the Uniteds did most likely own the actual gallery engine from the Mechanical and carried the name Old Lady to the "squirrel tail" suction engine presently on display at the Frederick County Fire and Rescue Museum.[60]

As the Mechanical Company also owned one gallery engine, two suction engines and two hose carriages, it is highly probable that the United Fire Company purchased two engines from the Mechanical, a suction engine and a gallery engine, one of which was called Old Lady.[61] The Baltimore Veteran Volunteer Firemen's Association was organized in 1877 with John Dukehart elected president; however, no M. Dukehart is listed in the history written by Clarence H. Forrest.[62] Exactly what or which Old Lady exists today will most likely remain a mystery for generations to come.

Chapter 3

EARLY FIRE PROTECTION IN FREDERICK COUNTY

Middletown Volunteer Fire Company

The town of Middletown was believed to be first occupied by German settlers in the 1740s, followed shortly thereafter by English settlers. The town was formally founded in 1767 when the first lots were sold.[63] The first documentation of early fire protection outside the city of Frederick was an act passed by the Maryland state legislature in 1803 to authorize a lottery to purchase a fire engine for Middletown. The lottery authorized up to $1,000; the purchase price of the engine was $400.[64] On January 13, 1825, a charter was authorized to incorporate the Union Fire Company of Middletown in Frederick County.[65]

Around 1850, the ladies of the community conducted a fair to raise money to purchase a new engine that was utilized by a "new" fire company organized taking the name of Independents. George Bowlus served as the president. A rival company organized as the Juniors and claimed the old engine.[66]

On July 13, 1851, a suction engine arrived in Frederick. This engine was purchased from the Patapsco Engine Company of Baltimore. The Patapsco Fire Company was organized in 1822. In 1843, the company had five pieces of apparatus, including one suction engine.[67] This is the engine believed sold to Middletown for a cost of $400. The total cost, including repairs and painting, was $500.[68]

An early engine protecting the town of Middletown. *Courtesy Middletown Volunteer Fire Company.*

Both Middletown companies were equipped with leather buckets, wooden ladders, fire hooks and eight hundred feet of leather hose. Ladders and buckets were prepositioned throughout the town to be handy in the event of a fire. The Independents had fifty-four members, and the Juniors had forty members. Both companies were uniformed. As was the case in many towns with multiple fire companies, fierce rivalries resulted in skirmishes between the two companies. Both companies were disbanded prior to the Civil War.

It was not until a major fire in Middletown in 1892 at the old tannery that the community realized the need for organized fire protection. The town board purchased five hundred feet of "rubber hose" that could be used on a fire plug, but no hydrant system existed. Instead of organizing a fire company, the town fathers initially took the first step toward improving fire protection by installing a water system. In 1893, the town began to install wooden water mains down Main Street with a fire hydrant installed every three hundred feet.

Though the town no longer had an organized fire company, the old 1850 engine was still maintained and was used to test the newly purchased hose. The engine was taken to "pools" of water. After testing the hose at the first pool, it was being moved to another pool down an incline when the engine, without any type of braking, became nearly unmanageable. One of the men slipped and was struck in the head by a suction pipe. The runaway engine just missed him as the wheels passed within six inches of his head.

In 1894, the burgess and commissioners of Middletown called a community meeting to discuss fire protection for the town with the intent of organizing a fire company. About sixty people expressed interest in meeting, and on January 26, 1894, the move for a fire company was inaugurated with thirty members secured. William V. Boileau, nearly seventy years old and a member of the former Independents, advised he could "make as good a run and hold a play pipe as effectively as the next one." The Boileau family continues to be active in the Middletown Volunteer Fire Company today. David Boileau was elected president for the year 2020 and tragically passed away before serving the term.

Millard Cook was elected president of the fire company in 1894. William G. Boileau was elected the first fire company secretary. The newly organized fire company would still use the engine purchased prior to the Civil War. The new fire company purchased a new hose reel, still owned and displayed by the Middletown company today. The company remained in service but was reorganized, at the request of the town burgess, George Doub. On March 6, 1923, the company serving the community today, the Middletown Volunteer Fire Company, was formally organized. The 1850 hand pumper was still the only engine serving the town. To update the fire protection, the newly reorganized fire company purchased a unique 1923 Kissel fire engine.[69]

Vigilant Hose Company, Emmitsburg

First settled in 1733, the town of Emmitsburg is located in the northernmost part of Frederick County and was officially founded in 1785. Early maps from the 1800s indicated an engine house was located on land roughly near 115 North Seton Avenue.[70] Like many very old communities, the use of a bucket brigade was most probably the first means of fighting fire. A notice issued on May 25, 1829, stated, "You, John Bader, are hereby requested to present yourself at the Engine House in Emmitsburg, on the 4th Saturday of May, June July and August, precisely at two o'clock p.m. with buckets and requisites for company training."[71]

The booklet published for the 125th anniversary of the Vigilant Hose Company provides an excellent vision of early fire protection in Emmitsburg. In 1840, an "out-dated" suction pump was purchased and a town ordinance

You — John Bader —
are hereby requested to attend at the engine
house in Emmitsburg on the 4th Saturday
of May, June, July, and August, precisely at
2. Oclock P.M., with buckets and other
requisites for Company training —

By the Pres't
May 25 1829
William Mooney Sec.

Training summons to John Bader to report for training at the Emmitsburg engine house on May 25, 1829. *Courtesy Vigilant Hose Company archives, photo by author.*

was passed requiring all men to join the fire company. Fines were levied for noncompliance. This action produced 141 members. Minutes from an October 4, 1851 town meeting noted that land for a new engine house could be purchased for $150 plus the payment of $12 in due taxes.

In 1851, a hand-pulled hose cart was built by a local Emmitsburg resident for a cost of thirteen dollars. The hose, made of riveted leather, is believed the first of this type used in the United States. A piece of this hose is preserved at the Smithsonian Institute. The hose cart built in Emmitsburg is an exquisite piece of firefighting history and is maintained and on display at the Vigilant Hose Company in Emmitsburg.

In 1884, a reservoir was constructed on the hills west of Emmitsburg. Water was piped into the town for the newly installed hydrant system. Also in 1884, the Vigilant Hose Company was organized, becoming a legal entity in 1910. The existing building housing the fire company was purchased in the 1920s for $4,500.

A hose cart built in Emmitsburg in 1851 responding to an alarm in Emmitsburg. *Courtesy Vigilant Hose Company.*

Charles Rowe, first president of the Vigilant Hose Company of Emmitsburg. *Courtesy Vigilant Hose Company.*

The Vigilant Hose Company 125th year anniversary history book notes several early fires in the town. In 1845, the Otter Hotel located on the southwest corner of the square burned. In 1848, the Elder and Taney Warehouse burned and nearly destroyed the Lutheran church.

On June 15, 1863, just two weeks before the Battle of Gettysburg, the town was nearly destroyed by fire. Fifty buildings northeast of the square, including the Western Maryland Hotel on the square, were destroyed. A week after the fire, Eli Smith, a resident of Emmitsburg and Union sympathizer, was arrested on suspicion of starting the fire to prevent advancing Confederates from taking supplies from the town. In July 1863, Smith was formally arraigned in Frederick and charged with arson.[72]

Though not officially part of the emergency services in Emmitsburg, the Sisters of Charity of St. Joseph's were noted for their nursing skills in the Civil War. During the Battle of Gettysburg just seven miles away, several of the sisters responded to assist in rendering aid to the wounded. Several hospitals were established in the town of Emmitsburg, with many of the Sisters of Charity providing the nursing care. The sisters were also summoned to the earlier battle at Antietam. In total, more than 280 of the Sisters of Charity provided medical aid to both Union and Confederate soldiers.[73]

Another major fire that had implications today occurred in the winter of 1885 at St. Joseph's College. Though newly formed Vigilant Hose Company responded and the town fire hydrants were functional, the blaze consumed the "new" building and threatened other structures on the campus. Help was requested from Hagerstown, Frederick and Baltimore. Only the arrival of help from Frederick has been noted in history. This fire will be highlighted in another chapter.[74]

The Vigilant Hose Company was formed in 1884. The organization became a legal entity in 1910. The fire company moved to the present location in the 1920s at a cost of $4,500. The first motorized apparatus was purchased in 1929 for $3,200. The Ford fire engine arrived in 1930 and was sold to the Rocky Ridge Volunteer Fire Company in 1951. The engine E131 was fully restored by the Rocky Ridge Volunteer Fire Company and is on display at Rocky Ridge.[75]

Libertytown Volunteer Fire Department

The unincorporated community of Libertytown, originally known as Dukes Woods, was first settled in 1782. The town was a bustling crossroads community that even had its own newspaper. The first fire company for Libertytown was chartered in 1847 as the Liberty Town Fire Company. The founding members were Thomas Sappington Jr., Francis S. Jones, Thomas Carr, Basil Sweadner, Enoch Beal, Ephraim Albaugh, Jonothan Browning, Henry Baker, Dennis Etzler, Simon Hartner and Andrew Arnold.

The corporation stated, "By the name and style of the Liberty Town Fire Company; the object of which company is hereby declared to be the protection of property of the citizens of liberty Town from destruction by fire." It was also noted in the charter that "the corporation of body politic shall not at any time hold or possess property, real, personal or mixed, exceeding in value of five thousand dollars."[76]

This particular version of the Libertytown Volunteer Fire Department apparently faltered, as on May 13, 1899, a disastrous fire struck the community and created a call to organize a fire company. A general store, feed store, tenant house, two dwellings and outbuildings were destroyed with an estimated $5,000 in damage.[77] An article in the May 20, 1899 *Daily News* stated a meeting was held to organize a fire department and to have the hand engine repaired.[78]

The engine so noted in the news article was the Old Lady described in an earlier chapter. At some point after the United Steam Fire Engine Company No. 3 purchased the Lily of the Swamp steam pumper, the Old Lady found its way to Libertytown. But the Liberty Fire Company apparently did exist in 1892, as the letter to initially organize the Maryland State Firemen's Association (MSFA) sent by the Independent Hose Company Committee for Arrangements for the 75th Anniversary also included Liberty Fire Company in the listing of Frederick County fire companies invited to help organized the MSFA.[79]

The fire company organized as the United Volunteer Fire Company, presumably taking the name already painted on the engine United. Taking the name already shown on fire equipment was a common practice, as was the case when the Water Witch Fire Company in Port Deposit bought a steam pump engine from Water Witch in Wilmington, Delaware, and assumed the fire company name already painted on the engine.[80]

An article in the August 18, 1899 *News* reported that the "old fire engine in Liberty has been overhauled and [is] now in readiness for service in case

Left: The original United Fire Company of Libertytown firehouse, abandoned in 1932. *Courtesy Frederick County Fire and Rescue Museum.*

Below: A 1915 Liberty Fire Company picnic with the believed first Black fire company member in Frederick County. *Courtesy Libertytown Volunteer Fire Department.*

of emergency. A fire company has been organized among the citizens of the town, and a feeling of more safety prevails now than heretofore."[81] The fire engine house was located on the north side of Main Street near the town creek.

This version of fire protection in Libertytown existed until at least 1915, as a picture dated August 22, 1915, shows the Liberty Fire Company picnic. As an interesting note, one of the members in the picture, identified as Charlie Key, is believed to have been the first African American member of a volunteer fire company in Frederick County and possibly the state of Maryland.[82] The exact date this fire company disbanded is unknown, but the engine was sold in 1932 to a George Conner to display in the museum of the Baltimore Equitable Insurance Company.[83] A 1935 picture of the old engine house on Main Street in Libertytown shows the structure very dilapidated and in disrepair. The spot where the sign "United Volunteer Fire Company" was hung could still be seen.[84] The present-day Libertytown Volunteer Fire Department was organized in 1964 and was the first of the modern-era fire companies in Frederick County to have African American charter members.

Guardian Hose Company, Mechanicstown/Thurmont

The original name of the town of Thurmont was Mechanicstown, but John Hood, president of the Western Maryland Railway, was required to change the name due to a duplicate and similarly named railway station. The name change was left to the citizens of the community, who chose the name Thurmont, most probably from the German *thur* meaning "gateway" and *mons* meaning "mountain," thus the meaning "gateway to the mountains." The town is truly the gateway to the mountain pass opened by the railroad.[85]

There is little documentation of the early fire protection in Thurmont or Mechanicstown. According to historian and present mayor John Kinnaird, the reason there was little early organized fire protection was attributed to the fact that early Thurmont never suffered a major fire. Many of the initial structures were brick and thus limited the spread of fire.[86]

The Guardian Hose Company was established in 1887. An 1888 photo displays a formal drill team with twenty-eight uniformed members. The first hose reel owned by the fire company is also displayed in the picture. On

July 3, 1890, a hook and ladder truck was received by the Guardian Hose Company. The truck was furnished by J.M. Kerrigan of Emmitsburg, and William H. Loy of Thurmont made the iron framing. The beams of the hook and ladder are lettered "Mechanicstown, Guardian Hose Company No. 1."[87]

In 1890, the Corporation of Mechanicstown requested the Maryland legislature consider approving the authorization of bonds to construct a "firemen's building and Town Hall." The bonds would pay 6 percent interest and run for a period of twenty years.[88] A building housing the town hall and the fire company was built on North Church Street in 1890. This building served as the quarters of the Guardian Hose Company until it was razed for a new fire station in 1950.[89]

A notice regarding the "ringing of the town hall bell" announces the signals of the Mechanicstown Fire Department would have been published sometime between the construction of the town hall in 1890 and the changing of the town name to Thurmont in 1893. The town was divided into three fire districts: First District, West Main Street: two taps; Second District, Water Street, Church Street, East Street and Boundary Avenue: three taps; Third District, East Main Street, Carroll Street and Lombard Street. A "General Alarm" could be sounded with a series of taps of ringing of the bell.[90]

The Guardian Hose Company was organized in 1887 and helped organize the Maryland State Firemen's Association at the organizational meeting in Frederick in 1892. Though participating in the early attempts

The 1888 "reel team" of the Guardian Hose Company of Mechanicstown. *Courtesy Frederick County Fire and Rescue Museum, John Kinniard collection.*

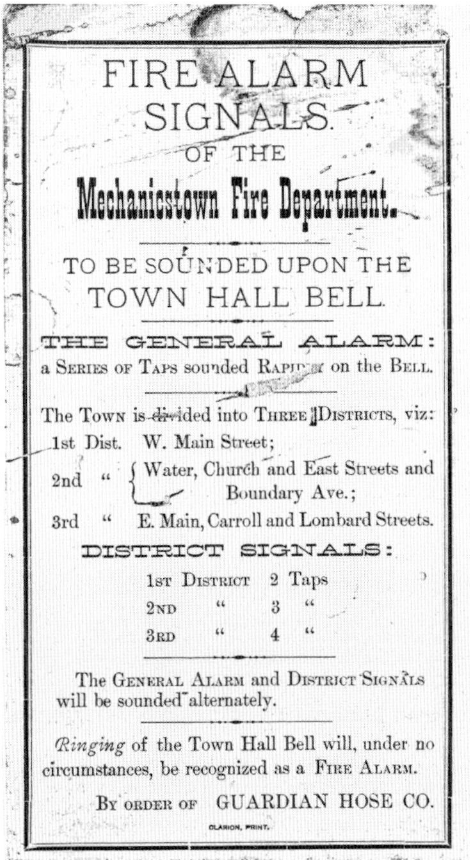

FIRE ALARM
SIGNALS
OF THE
Mechanicstown Fire Department.

TO BE SOUNDED UPON THE
TOWN HALL BELL.

THE GENERAL ALARM:
a Series of Taps sounded Rapidly on the Bell.

The Town is divided into Three Districts, viz:

1st Dist.	W. Main Street;
2nd "	Water, Church and East Streets and Boundary Ave.;
3rd "	E. Main, Carroll and Lombard Streets.

DISTRICT SIGNALS:

1st District	2 Taps
2nd "	3 "
3rd "	4 "

The General Alarm and District Signals will be sounded alternately.

Ringing of the Town Hall Bell will, under no circumstances, be recognized as a Fire Alarm.

By order of GUARDIAN HOSE CO.

Clarion, Print.

Left: Notice of "Fire Alarm Signals of the Mechanicstown Fire Department" by order of Guardian Hose Company, circa 1890. *Courtesy Frederick County Fire and Rescue Museum, John Kinniard collection.*

Below: Remains of the Excelsior Mill fire that killed Guardian Hose Company firefighter Harry C. Root on June 18, 1898. *Courtesy Frederick County Fire and Rescue Museum, John Kinniard collection.*

to organize the MSFA, the Guardian Hose Company did not join the MSFA until 1928.[91]

On Saturday, June 18, 1898, Harry C. Root of the Guardian Hose Company became the second line-of-duty death of a firefighter in Frederick County. At approximately 2:00 a.m. on June 18, a fire destroyed a building of the Crescite Excelsior Works in Thurmont. The fire had great headway by the time the Guardian Hose Company arrived, but the firefighters were able to save the office. While fighting the fire, an iron smokestack fell among the firemen and instantly killed Root. As of this writing, this tragedy is the only firefighter line-of-duty death to occur in Frederick County as a result of injuries sustained in firefighting operations.[92]

Brunswick Volunteer Fire Company

The city of Brunswick was originally settled in 1728. A grant to the land was then given to John Hawkins by George II of Great Britain on August 10, 1753. The land was sold, and Leonard Smith platted it in 1787 with the name of Berlin, as many Germans settled in the area. The town grew from the initial plat along the river to the many hills that extended from the Potomac. The name Berlin, however, could not be used for mail, as there was already a Berlin on the Eastern Shore of Maryland. The town continued to grow and was incorporated on April 18, 1890, with the name Brunswick. The year 1883 changed the town forever with the building of a massive rail yard by the Baltimore and Ohio Railroad.[93]

The earliest form of fire protection in Brunswick was provided by citizen bucket brigades. In 1890, the City of Brunswick instituted "water companies" that were also to be utilized for fire protection. This included a central water system and fire plugs. The City of Brunswick installed "reel houses" throughout various sections of the town. Some of the reel houses had a standpipe that received water from a well at the locations. The first reel consisted of a rod between the hubs of two large wheels with the hose wrapped around the cross rod.

A reel house was located in each of the three wards of the city. One reel house was located on the south side of Potomac Street opposite First Avenue. A "fire wagon" was also located at this early firehouse. This station had to be built on stilts due to the steep terrain. Another reel house was built on New York Hill on the east side of Ninth Avenue. This station had a stand

A Brunswick Wirt-Know hose reel restored and owned by Clarence Weber, life member of the Brunswick Volunteer Fire Company. *Photo by Chief George Clary.*

pipe configuration to supply water. A third station was located on Wenner's Hill at the corner of Maple Avenue and F Street. Firefighting water was supplied by a water tower that was built on steel framework.

A fourth reel house was built in the 600 block of West Potomac Street. The reel houses along Potomac Street utilized water from fire plugs that were located downtown. Water for these hydrants was gravity-fed with adequate pressure for firefighting. Each reel house had a chief. Each station had a glass-covered hole with a key so anyone could unlock the door to the fire equipment. One reel was a large nine-man reel, and the other three smaller reels were three-man reels.

In 1897, an organized fire protection unit in Brunswick was formed. The equipment included a Howe hand pumper, a hook and ladder wagon and hose reels. Horses owned by the B.P. Crampton Company would be used to pull the equipment to fires outside the town. The first Brunswick Volunteer Fire Company was formed on May 10, 1910, with Edward C. Shaffer as president. Firemen's Hall on West Potomac Street was erected under the supervision of Lorenzo S. Gardner.[94]

Author's note: The Baltimore and Ohio Railroad also maintained "fire stations" throughout the railroad yard. These red buildings housed hoses, extinguishers and other items that could be used for firefighting. A firefighting hose reel was kept in the roundhouse as well.

Jefferson Volunteer Fire Company

The earliest known effort for fire protection in the town of Jefferson was most likely bucket brigades. In 1919, two hand-drawn Badger chemical trucks were purchased by the community. George Kolb of Frederick demonstrated the firefighting capability of the two chemical engines. Each hand-drawn chemical engine had a capacity of forty gallons. One unit was located in each end of town.[95] The two chemical engines were pressurized with soda/acid, much like a fire extinguisher. One of those units has been restored and is now on display at the Frederick County Fire and Rescue Museum.

A 1919 Badger chemical engine, one of two that protected the town of Jefferson. It is on display at the Frederick County Fire and Rescue Museum. *Photo by author.*

In 1955, the Braddock Heights Volunteer Fire Company established a substation in Jefferson. For many years, this fire station was known as Jefferson Station and was dispatched as such. Braddock Heights purchased a 1948 Ford and placed that unit at the Jefferson Station.[96] On Saturday, April 12, 1969, a fire broke out in the fire station. It was believed to have started in the wiring of the cab of a 1947 Chevrolet engine that was also placed in the Jefferson Station by Braddock Heights. The fire destroyed the 1947 Chevrolet engine, with some fire damage and heavy smoke to the fire station. The 1948 Ford was pulled out of the firehouse undamaged and was used to fight the fire until arrival of other area fire companies. Fortunately, the 1957 Diamond-T engine that was purchased from Berwyn Heights Fire Department to replace the destroyed unit was being painted at another location at the time of the fire.[97]

A public meeting was held on April 6, 1970, to begin discussions of separating from the Braddock Heights Volunteer Fire Company and forming a separate fire company in Jefferson. On Thursday, June 4, 1970, the firefighters serving in the Jefferson Station submitted an application to the Frederick County Volunteer Firemen's Association to organize a separate fire company. The Jefferson Volunteer Fire Company went in service shortly thereafter and was assigned the number Company 20.[98]

Walkersville Volunteer Fire Company

Fire protection for the town of Walkersville began in 1909 with the formation of the Walkersville Water Company. In coordination with installation of water lines and hydrants, the Walkersville Hose Company was organized with hand-drawn hose reel. These actions most likely took place after a fire destroyed the residence of John Jamison. Only a bucket brigade and fire extinguishers from the Monocacy Valley Canning Company factory were used to fight the fire.[99] The minutes of December 12, 1927, of the Walkersville Hose Company (unincorporated) show that T.R. Saylor was president and captain was Clyde Fogle. A reorganization was effected in 1939, and the result in 1940 was the incorporation of the Walkersville Volunteer Fire Company, Inc. During incorporation in 1940, Saylor was elected president, and fire chief was Luther Horine Sr.

The first carnival was held in 1940 to raise funds for a motorized fire engine. The net profit was approximately $750. The company purchased a new engine through Renner Motor Company in New Midway. The company bought a 1940 Ford Howe pumper carrying three hundred gallons of water at a cost of $4,700. The original 1940 Ford fire engine has been restored in brilliant Walkersville Blue and is on display in a special room in the front of the fire hall. The tradition of Walkersville Blue fire apparatus continues from 1940 until today.[100] Though there is no documentation as to the reason the fire engines have been painted blue, Walkersville life member Vaughn Zimmerman has noted the Walkersville Lions Club was instrumental in the early organization of the fire company, and the fire company originally met at the Walkersville High School before they owned their own building. Both the Lions Club and the Walkersville High School colors are blue and gold, which may explain the unique fire engine colors.[101]

Braddock Heights Volunteer Fire Company

The community of Braddock Heights is unique, as the village was built as a vacation resort by the Frederick Railroad (later Hagerstown and Frederick Railway). Though today it is known as Braddock Mountain, the original name of the mountain was Catoctin Mountain. The owners of the railroad and developers of the park and resort named the area Braddock Heights in honor of General Braddock, a name that would be familiar to those in Washington, D.C., whom the railroad hoped to attract to purchase land.[102]

As the community was planned, so was the fire protection. By 1910, most of the buildings constructed in the previous two years were occupied. E.S. Mobley and Brothers were hired to install fire hydrants in the resort area to be used by the "new Braddock Fire Department."[103]

The Braddock Heights Fire Department got its first official home on October 15, 1912. The building was constructed by Frederick Railroad Company. It was used to store equipment and house a substation on the corner of Maryland Avenue and Jefferson Boulevard. On August 8, 1935, the fire department put a 1923 floodlight truck in service as the first piece of motorized apparatus owned by the fire company.[104] This unit was most likely the Dodge/Waterous floodlight truck formerly owned by the United Fire Company, as the model year and the year the Uniteds Diamond-T rescue truck went into service correspond.

Chapter 4

FIREMEN SOLDIERS AND JOHN BROWN'S RAID ON HARPERS FERRY

In January 1858, a public hanging was held on Cannon Hill with no local militia unit to guard the gallows. This lack of a local militia led to the birth of the United Guard, a local militia composed primarily of members of the United Fire Company and informally associated with the fire company. A week later, the United Guard was inspected by Colonel Shriver, commander of the Sixteenth Maryland Regiment of the Maryland National Guard, to which the local unit became attached. On April 5, 1858, forty-six uniformed members of the United Guard paraded through the streets of Frederick led by the American Band, an organization also associated with the United Fire Company. On June 1, 1858, the United Guard offered to serve as marines on ships of war; the offer was appreciated but respectfully declined by the War Department.[105]

The two other fire companies in Frederick soon followed and marshaled together a militia unit. The Junior Fire Company organized the Junior Defenders under the command of Captain John Ritchie. Later, the Junior Defenders came under the leadership of Captain William Glessner, and the company was known as Company B. The Juniors detachment went under federal jurisdiction and came under the command of Colonel William P. Maulsby. This move placed the unit as part of the First Maryland Potomac Home Brigade (PHB). The Independent Hose Company organized the Independent Rifles (or Riflemen) under the direction of Captain Ulysses Hobbs. Ultimately, the fire company militia units were part of the

A hat with Independent Riflemen "IR" and epaulets worn by David T. Bennett, an officer in the United Guard. *Displayed at United Fire Company, photo by author.*

Maryland Militia that would later become the beginnings of the Maryland National Guard.[106]

On June 19, 1859, the United Guard led by Captain J. Thomas Sinn and Junior Defenders led by Captain John Ritchie paraded in full uniform with the American Band. It was noted the United Guard received weapons of "Minnie rifles and muskets" in July 1859.[107]

John Brown was an abolitionist who served as a conductor on the Underground Railroad and established the League of Gileadites, an organization to help runaway slaves escape to Canada.[108] Brown continued his abolitionist pursuits with his sons in the Kansas Territory. Several days after a raid on Lawrence, Kansas, by proslavery activists, Brown led an armed party of men to Pottawatomie Creek to attack men he believed to be proslavery, dragging five men and boys from their home and murdering them. He also conducted a raid in Missouri, freeing eleven slaves and killing the slave owner. His actions helped ignite the internal statewide struggles between proslavery and abolitionists that became known as Bleeding Kansas.[109]

John Brown moved east under the alias of Isaac Smith. He arrived in Harpers Ferry, Virginia (now West Virginia), with his sons Oliver and

Owen and Jeremiah Anderson and moved into the farmhouse on Samples Manor, the Kennedy Farm near Knoxville, Maryland. Brown planned to raid the federal arsenal at Harpers Ferry, anticipating slaves would join his force. He would be able to arm the slaves with weapons from the arsenal and lead a revolt.

On the night of October 16, 1859, John Brown put his plan into motion, raiding the town and the federal arsenal in Harpers Ferry. At approximately 10:00 p.m., his raiders took control of both bridges, the U.S. Armory and Arsenal and the U.S. Rifle Works on Hall's Island. Two men—slave owners Lewis Washington and John Allstadt—were taken hostage and their slaves freed. Brown held a Baltimore and Ohio passenger train at the Harpers Ferry station. Ironically, the first man killed during the raid was Hayward Shepherd, a free Black man working for the Baltimore and Ohio Railroad.[110]

On October 17 at around 8:00 a.m., a train arrived at Monocacy Junction (now Frederick Junction) reporting an insurrection underway in the town of Harpers Ferry. Stationmaster John T. Quynn sounded the alarm in Frederick.[111] Upon receiving this information, the Frederick militia units immediately offered their services to President James Buchanan, who accepted the offer. Arrangements were made for an immediate departure. This was the first military offer of assistance received by the president.[112]

"Harpers Ferry Riot—The Independent Bell & United or Swamp Bell are both now ringing (Swamp first) calling together the military companies of our city. The report from Harpers Ferry is that there is a kind of insurrection among the Negros of Jefferson County Virginia & want to seize the United States arms there or that the workmen at the Ferry are somewhat rebellious. Monday October 17, 1859 10 o'clock AM." This entry in the diary of Jacob Engelbrecht provides a vivid image of how the volunteer fire companies of Frederick summoned the militia members. "Monday, October 17 ½ past 3 o'clock PM Our three town companies Captain Sinn, Captain Ritchie & Captain Hobbs just started for the cars at the Depot for the scene of War—Harpers Ferry."[113] The three companies mustered approximately 175 total members and requested a cannon owned by Frederick city be transported by the earliest train.[114]

Citizens of the town stormed the U.S. Rifle Works. Harpers Ferry mayor Fontaine Beckham was shot and killed by Brown's men. Brown's men fled into the armory firehouse, where he and his raiders barricaded themselves. The building was built in 1848 to house the military arsenal fire engine and serve as a guard house; thus, it served as a sturdy building to hold off attack. Marines led by Captain Robert E. Lee and Lieutenant J.E.B. Stuart arrived

Baltimore and Ohio Railroad passenger station located across from the United Fire Company where Frederick militia units departed for Harpers Ferry. *Courtesy Frederick County Fire and Rescue Museum.*

in Harpers Ferry on October 17. On the morning of October 18, John Brown refused to surrender to Lieutenant Stuart. Ten of John Brown's men were killed in the raid, as well as six civilians.[115]

Initially, the Frederick companies were asked to form a battalion to assist in the storming of the engine house. Privates Sinn, Sencil and A. Boteler under Captain John Sinn of the United Guards; B.T. Johnson, Mulhorn and Smith under Captain Ritchie of the Junior Defenders; and Rarster, Simpson, Buckey and Frazier under Captain Hobbs of the Independent Rifles were mustered to participate to help storm the engine house where John Brown and his men were holing up. However, after waiting a considerable period of time, Captain Robert E. Lee determined only the marines would storm the rebellious party. The marines stormed the engine house and captured Brown and his party. The men were held as prisoners in the engine house until they were to be transported to Charles Town, Virginia (now West Virginia), for trial.[116]

Marines, as well as military units from Baltimore, Charles Town, Winchester and Shepherdstown, also responded to the call to arms. Upon arrival, the Frederick fire company militia units were initially assigned to guard the railroad bridges leading into Harpers Ferry.[117] Captain Sinn and members of the United Guard were also assigned the duty to guard the engine house where John Brown and his men were being held. Brown spoke to Captain Sinn to try to negotiate a surrender. Brown tried to convince Captain Sinn to allow the raiders to leave Harpers Ferry via the bridge to Maryland, where Brown and his men could "take their chance for their lives in an open fight." Captain Sinn refused to release Brown or any of his men.

However, Brown complained that his men were "shot down like dogs" while carrying a flag of truce.[118] Captain Sinn was concerned about medical treatment of the wounded and requested Dr. William Tyler of the Junior Defenders to assist. However, the United States Marines prevented any medical aid to Brown or his men. Dr. Tyler did accompany marines to the wounded military and rendered aid to Private Quinn, who was fatally wounded, and to Private Rupert, who was also wounded but survived. Dr. V. Delashmutt with the Independent Rifles and Dr. John Goldsborough with the United Guards also accompanied the fire companies to Harpers Ferry.[119]

On October 18, 1859, the home guard fire company units returned to Frederick carrying John Brown's pikes. Upon the return, the fire company militia units were assigned additional duties to guard the railroad bridge at Monocacy Junction and guard the federal arsenal then located on the property of what today is the Maryland School for the Deaf. The United Guard was also summoned back to Harpers Ferry later that year to guard

John Brown and his men hold hostages inside the Harpers Ferry engine house just before the military makes entry. *Library of Congress, www.loc.gov/item/2002735881.*

the railroad bridge. The Maryland legislature authorized payment to each fire company home guard of $200 for their service at Harpers Ferry. The United Guard received an additional $200 for the return assignment.[120]

The trial of John Brown and his men was held in Charles Town at the end of October and into November. Brown was found guilty of "high treason and murder in the first degree." Several other members of his party were found guilty of various levels of crime. During the trial, Captain John Sinn testified on behalf of John Brown as a witness. He advised that Brown complained to him that he and his men were "shot down like dogs" while carrying the flag of truce and that the abolitionists had shot at no one who had not fired on them first. Sinn also testified he saw George B. Shope of Frederick being held as a hostage by Brown. Aaron D. Stephens, a member of John Brown's raiders, was wounded and held by some local men in a hotel. Sinn advised he probably saved the man's life, as he would have been shot while he lay in bed.[121]

Brown's fellow conspirators, John Cook, Edwin Coppie, Shields Green and John Copeland, were also found guilty. All were sentenced to death by hanging. John Brown was hanged on December 2, 1859. The other men were hanged on December 16. Over one hundred rifles, pistols, kegs of powder and a large assortment of armaments were discovered at the Kennedy farm, used by John Brown to execute the Harpers Ferry raid.[122]

Chapter 5

THE CIVIL WAR YEARS

Early in the year 1860, several military units began to form across the state. Two of the earliest units in Frederick County were the Liberty Riflemen with Dr. A.A. Sappington, a familiar family in the Libertytown fire service, and the Union Cavalry Troop of Unionville under the direction of Captain M.C. McKinstrey. Governor Hicks appointed leadership in the state militia. As the Sixteenth Regiment was formed, Edward Shriver was appointed colonel and James McSherry lieutenant colonel. Both men served with the Independent Hose Company. In August 1860, the Maryland legislature voted to provide funds for the uniforms of the Independent Riflemen. The uniforms were a dark green frock coat, single breasted, with three rows of regulation "R" buttons and trimmed in gold lace.[123]

The fire company militia units continued to function into the early days of the Civil War. The Junior Defenders guarded the arsenal at what is now the Maryland School for the Deaf and the Monocacy railroad bridge. The Junior Defenders also conducted a ball for "gentlemen and ladies" on April 9, 1860. On April 24, 1861, Major General Kimmel assigned three fire company militia units to guard the armaments of barracks. By then, the leadership of the units had somewhat changed: Captain Ritchie continued to lead the Junior Defenders, Captain Beckley was leading the United Guards and Captain Saunders led the Independent Riflemen. The Zouave Cadets were also detailed to guard duty under the leadership of Captain Baugher.[124]

As the Civil War broke out, Frederick was truly a community split between the Union and Confederacy. Fire company membership became split, with supporters for both sides. The split allegiance of fire company members most likely attributed to the demise of the fire company militia units and the advent of home guards. In May 1861, the Junior Defenders disbanded, as some members joined the Potomac Home Brigade.[125] The fire company militia units probably meshed into various local military groups. An entry in Engelbrecht's diary on Monday, April 26, 1861, notes the formation of Brengle's Home Guards. A further review of the members of this local militia unit shows many names associated with the three volunteer fire companies.[126] Family members of several names mentioned in the list of the Brengle's Home Guards continued to serve in the Frederick volunteer fire companies into the 1900s.

The Frederick Zouaves were mustered on September 18, 1861, as part of the First Regiment of the Home Brigade. The extensive list of names noted as Zouave members in the diary includes several members of the various fire company militia units.[127] Zouave units organized by volunteer fire companies were common in the Civil War. Zouave fighters were originally fierce Algerian fighters prior to the French occupation of 1830. The Zouave fighters were noted for their bright colors, ornate uniforms and dashing appearance. These fighters were recruited into the Crimean conflict by the French military; ultimately, Zouaves were French.[128]

The existence of a Frederick Zouave unit would provide logical reasoning for the more ornate uniforms for fire company members such as those described for the Independent Hose Company and the bright red military-style uniform with ornate fez hat on display at the Junior Fire Company. A picture of an unknown member of the Junior Fire Company appears to also show an ornate military uniform similar in style to a displayed uniform at the Junior Fire Company, noted as a Junior Defender.[129]

Captain William Glessner, a member of the Junior Defenders at Harpers Ferry and later a captain of the Junior Defenders, also organized a company initially called Union Defenders of Frederick that was mustered into action on September 6, 1861.[130] Company B, as it became known, came under the federal jurisdiction of Colonel William P. Maulsby, First Maryland Potomac Home Brigade.[131]

On September 28, 1861, 101 members of the 114th Pennsylvania Infantry Regiment Zouaves d'Afrique, Collis Zouaves, arrived in Frederick by train. Engelbrecht noted, "Their dress consists of blue doublet with yellow braid and red pantaloons, white canvass leggings & shoes & red 'fez.'

Left: An unknown member of the Junior Defenders. *Courtesy Jim and Joy Deater.*

Below: A re-creation of the parade uniform of a Junior Defender on display at the Junior Fire Company No. 2. *Photo by author.*

Their appearance very prepossessing that is, a handsome dress."[132] The Zouaves of Captain Charles Collis returned to Frederick on December 4, 1861, to serve as provost guards of the town. The unit assumed quarters in the upstairs of Coppersmith's Hall.[133]

During the Civil War, Frederick had citizen sympathies for both the Union and the Confederacy. After Fort Sumter was fired on, soldiers were called to defend Washington. As the regiment was passing through Baltimore, they were attacked by a mob in an event that became known as the Pratt Street Riots. Four soldiers from the Sixth Massachusetts and eleven rioters were killed in the assault.[134]

As a result of this riot and the strong Southern sympathies around Baltimore and the Chesapeake Bay area, Governor Hicks decided to convene the legislature in Frederick, Maryland, a more pro-Union city. The General Assembly initially attempted to meet on April 26, 1861, in the Frederick County Courthouse. However, as the building was too small, the General Assembly moved to Kemp Hall, a building owned by the German Reformed Church, today the Evangelical Reformed United Church of Christ.[135] On April 30, the weekly *Frederick Herald* reported, "The Legislature seems comfortable and well provided for in their new halls in the German Reformed Building."

The original Frederick County Courthouse was built in 1756. The second courthouse was built in 1857. In 1842, a major fire struck the courthouse as flying embers from a burning chimney on a house on Court House Square set several houses on fire, eventually spreading to the cupola of the courthouse.[136] Frederick had several fire companies in existence at the time of the fire: Independent Hose Company, Junior Fire Company, Washington Hose Company, Friendship, Sunrise and Auckerman. However, only Independent and Junior had pumping engines.[137]

The Maryland General Assembly met from April 26, 1861, until May 14, 1861. The legislature adjourned until June 14, 1861, to again meet in Frederick. In the early morning of May 8, between 1:00 a.m. and 2:00 a.m., the courthouse was "set on fire & burnt nearly out, except the lower part where court, lawyers & jury usually sat." The entire roof was on fire when first discovered.[138] According to a report in the *Frederick Examiner*, "The books and papers in the offices of the Sheriff, Tax Collector, and Clerk to the County Commissioners were all saved," but a large volume of proceedings and other records were destroyed. Within a week of the fire, the legislature authorized a $40,000 bond issue to finance a new courthouse.[139] The Mutual Insurance Company of Frederick County insured the loss in

A painting of one of the fires of the Frederick County Courthouse. *Collection of author, courtesy Heritage Frederick.*

the amount of $5,000.[140] When this fire occurred, several engines were available from the Independent Hose Company, Junior Fire Company and the United Fire Company.

No specific cause of the fire was determined, though "Rebel rascals" were suspected.[141] The Frederick County Jail burned the following year on September, 13, 1862.[142] During the Civil War, the United Swamp Hall was used by the Federal government, and in 1875, the government paid the Uniteds $168 for the use.[143] The Junior Hall was utilized as part of General Hospital #6, which was a series of buildings including Ashbury Church, Primary Schools 71 and 72 and Junior Hall. General Hospital #6 operated from September 14, 1862, to February 11, 1863, treating a total of 497 patients.[144] An entry in Engelbrecht's diary on September 30, 1862, noted the number of wounded soldiers coming to Frederick hospitals from the Battle of Antietam: "Still we have in our city from 5 to 6000 and they are bringing from 4 to 600 every day. About 100 ambulances arrive every day (they make one trip a day 22 miles) and all are full."[145]

The second location of the Junior Fire Company at Market and Second Streets. *Collection of author.*

In 1862, the upper hall of the Juniors was also utilized to confine Confederate prisoners of war and as the office of the provost guard. The Sons of Temperance were also an occupant of the building and requested payment from the Federal government for damages to the property during the use of their section of the building. Much of the furniture of the organization was missing or damaged. In a letter to the quartermaster on February 10, 1864, sent by Isaac P. Suman and John C. Hardt, trustees, the damage sustained was estimated at $305.[146]

In September 1862, Brooklyn, New York firefighters who were members of the Fourteenth New York State Militia, Eighty-Fourth New York Volunteer Infantry Regiment, participated in the Battle of South Mountain. As they marched toward South Mountain west of Middletown, they found the bridge over Catoctin Creek had been set on fire by the Confederates attempting to slow Union advance. As the Union troops moved west, the members of the Brooklyn Fourteenth who were Brooklyn firefighters were found using a Middletown fire engine to extinguish the burning bridge.[147]

In 1868, a major flood of Carroll Creek occurred, damaging many homes and businesses in downtown Frederick. The flood began on the morning of Friday, July 24, 1868. It was believed to be the greatest flood of the town since its founding, as Engelbrecht noted in his diary: "Out-tipts every high water since the town was laid out." It was noted that houses partially collapsed, and some nearly washed away. Houses were inundated on Market Street.[148] The records of the United Fire Company were lost in the flood when the home of fire company secretary O.S.J. Beckley was destroyed.[149] Coincidentally, major floods of Carroll Creek occurred on July 24, 1821, July 24, 1848, and July 24, 1868.[150] The Swamp Bell of the United Fire Company would be sounded to warn townspeople of the rising creek.[151]

Chapter 6

JUNIOR STEAM FIRE ENGINE COMPANY NO. 2

FIRST STEAM FIRE ENGINE

As firefighting technology advanced, steam-powered fire engines began to replace the hand pumpers that were the standard fire protection of the day. A patent for the steam engine was awarded in England to Thomas Savery in 1698. In 1841, the first steam fire engine was tested in New York City. The first steam fire engine in Maryland was Alpha in the city of Baltimore in 1858.[152] The first steam fire engine in Frederick was purchased by the Junior Fire Company in 1876. The engine was built by the Silsby Manufacturing Company of Seneca Falls, New York.[153]

On September 20, 1876, two steam fire engines arrived in Frederick to be tried and tested by the Junior Fire Company in anticipation of purchasing a steam-powered pumper. An engine from the Silsby Manufacturing Company and the LaFrance Fire Engine Company of Elmira, New York, were tested at various areas around Frederick with other fire companies.[154] The Junior Fire Company purchased the Silsby engine at the cost of $4,000, including a complement of five hundred feet of hose. The engine was a solid wheeled engine, initially pulled by company members. On January 21, 1877, the Independent Hose Company voted to purchase a steam pumper from the LaFrance Fire Engine Company.[155]

In 1878, the Junior Fire Company petitioned the Maryland General Assembly to change the fire company name to incorporate "Steam" and "Number Two" into the official corporate name.

Laws of Maryland.

Chapter 33
AN ACT to amend an act passed at the December session of the year eighteen hundred and thirty-nine, chapter three hundred and three, entitled "An act to incorporate the Junior Fire Company of Frederick city, in Frederick county," and the supplements thereto, by changing the name of said company.

SECTION 1. Be it enacted by the General Assembly of Maryland, That the name of "The Junior Fire Company of Frederick," incorporated by the act of the December session, eighteen hundred and thirty-nine, chapter three hundred and three, entitled "An act to incorporate the Junior Fire Company of Frederick city in Frederick county," and the supplements thereto, be and the same is hereby changed to the name of "The Junior Steam Fire Engine Company, Number Two, of Frederick," and that the said "Junior Fire Company of Frederick" shall be hereafter known and designated by the name of "The Junior Steam Fire Engine Company, Number Two, of Frederick," and shall, under the name last aforesaid, have and succeed to all the rights, privileges, powers and property of the said "The Junior Fire Company of Frederick," and shall be subject to all the debts, engagements, contracts and liabilities of the same last-named company, it being the intent of this act to effect no other change in the said "The Junior Fire Company of Frederick" except a change in the name thereof.
Approved February 14, 1878[156]

Steam pumping engines only carried "hard" suction sleeves and were normally accompanied to the scene of a fire with a hose reel or hose wagon. Over 130 years after the Silsby engine was purchased by the Juniors, what is believed to be the accompanying hand-drawn hose reel was located in Frederick. The Junior Fire Company No. 2 proudly owns that hose reel. Jim Deater, chairman of the Historical and Archives Committee of the Juniors, has taken this reel under his wing by cleaning and repairing the wooden dry rot. In the future, it is hoped to restore the hose reel to its original form, including the mounting of the name "Junior" atop the metal arch. This hose reel most likely served until the purchase of a hand reel and hose carriage in 1889.[157]

The first actual hook and ladder apparatus in Frederick was purchased by the Junior Fire Company in 1877. The hand-drawn hook and ladder was

Left: The first steam pumper in Frederick, an 1876 Silsby steam engine of Junior Fire Company. *Collection of author.*

Below: A Silsby hose reel believed to be purchased in 1876 by the Junior Fire Company. *Photo by author.*

built by famed apparatus builder Charles T. Holloway. Holloway was the first fire chief of the Baltimore Fire Department and the Baltimore County Fire Department. He was a noted apparatus builder on the East Coast. Noted later in this book, he also served as an early officer of the Maryland State Firemen's Association. This apparatus was later modified to accommodate horses and rebuilt in 1913 with stronger ladders. In June 1889, the Junior Fire Company received a new hose carriage.[158] Also in 1889, the fire company installed steam heat in the building.[159]

In February 1903, the Junior Fire Company appointed a standing committee at the recommendation of Foreman William G. Shipley to study the possibility of horses being utilized for pulling apparatus to fires.[160] On April 25, 1903, a report ran in the *News* titled "Better Protection Against Fire." The article highlighted that there was no longer sufficient manpower to pull the engines as the "desire to 'run with the engine' has to a large extent died out." The article noted that often only a few men were pulling the engine while hundreds stood along the streets watching.[161]

The committee of the Juniors—consisting of Judge John Motter, Danial H. Steiner, John Eisenhauer, Frank Sappington, William H. Shipley and William Crimmins—traveled to Baltimore and visited fourteen engine houses to get ideas on remodeling the engine house and accommodating horses. They were hosted by Captain William Dunn of the Baltimore Fire Department. The building was remodeled, and on June 2, 1903, two gray horses, Dan and Billy, were put into service. A paid driver was also hired to be on duty twenty-four hours a day.[162]

In 1904, the Junior Fire Company purchased an American LaFrance combination chemical and hose wagon, delivered to the company on November 21, 1904. This majestic unit was painted maroon and black and was a horse-drawn wagon. The unit carried a thirty-five-gallon chemical extinguisher and two hundred feet of chemical hose. The unit also served as a hose wagon for the steam pumper and carried one thousand feet of firefighting hose. The hose and chemical wagon was built by the American LaFrance Fire Engine Company for a cost of $2,800. American LaFrance went on to become a premier builder of fire apparatus, and the Junior Fire Company would go on to purchase more equipment from the company into the motorized age.[163]

This dependable apparatus served the Junior Fire Company well into the motorized era. In 1913, the company placed the American LaFrance chemical wagon body on a White motorized vehicle chassis. The new chassis cost $3,500 and was delivered in thirty days. The vehicle had a thirty-

Junior Fire Company 1904 LaFrance combination chemical and hose wagon. *Courtesy of Sherry Kemp.*

1935 Ford-first piece of motorized equipment of the Funkstown Fire Co.

This 1935 Ford with the body of Junior 1904 LaFrance named Nellie Belle became the first fire engine of the Funkstown Volunteer Fire Company. *Collection of author.*

Driver "Uncle Billy" Davis and firefighter Glessner with a 1908 LaFrance engine. *Courtesy Sherry Kemp.*

horsepower engine and electric starter with a top speed of forty-five miles per hour. The refurbished chemical and hose wagon also received a new fifty-gallon chemical tank and one thousand feet of new hose.[164]

The life of the unit did not stop there. In 1931, the motorized chemical and hose wagon was sold to the newly organized Funkstown Volunteer Fire Company as that fire company's first piece of motorized apparatus. In 1935, the Funkstown company purchased a Ford chassis from the Hebb Motor Company of Funkstown and separately purchased a 350-gallon-per-minute Darley pump. Utilizing the original American LaFrance body and equipment, the fire company constructed its first fire pumper, nicknamed Nellie Belle.[165]

The American LaFrance Fire Engine Company was again contacted when it became time to replace the thirty-two-year-old Silsby steamer. In 1908, the Junior Steam Fire Engine Company No. 2, as the corporate name change officially read, purchased a combination Metropolitan steam engine from American LaFrance for a cost of $4,500. This was a heavier engine than the original Silsby.[166] On Saturday, September 12, 1908, George William Shipley, foreman of the Junior Fire Company, witnessed the testing of the new Metropolitan fire engine in Hagerstown.[167]

Chapter 7
TRAGEDY AND CHANGE

LINE-OF-DUTY DEATH OF A FIREFIGHTER

Tragedy struck the Junior Fire Company on Sunday morning, March 20, 1910. An alarm of fire was received for a fire in a storeroom on West Patrick Street. William "Uncle Billy" Davis was hired as a paid driver when the Juniors received horses in 1903. He was formerly a police officer under Mayor Chilton's administration. One of the original Junior horses was disabled and removed from service, and a new horse had taken its place a few months prior. When the alarm was received, driver Davis first had to move the hose and chemical wagon that was situated in front of the steamer out of the building so he could respond with the American LaFrance steam engine.

The article in the *Frederick News* on Monday, March 21, 1910, reported that with the horses hooked up, the new horse "excitedly plunged forward and Mr. Davis, endeavoring to stop it, was crushed against the wall and dragged forward. At the doorway he struck against an iron bar which is part of the apparatus for opening and shutting the doors and was thrown beneath the engine's wheels just as it emerged upon the sidewalk crossing. The four-ton engine passed over his chest, crushing the life out of him. The horses were caught and stopped by Officer Stull of the police force."[168]

Driver Davis was sixty-five years of age and an original paid driver of the Junior Steam Fire Company No. 2. He was the grandfather of Tom Fox, president of the Junior Fire Company No. 2 from 1988 to 1991, and of Lois Spurrier, wife of W.M. Ray Spurrier, an active member and officer of the

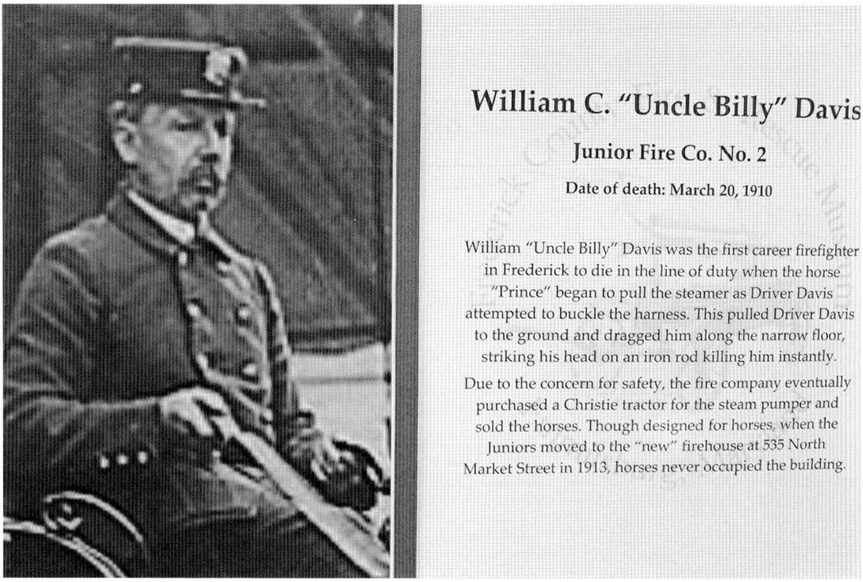

Memorial to William "Uncle Billy" Davis at the Frederick County Fire and Rescue Museum Hall of Honor. *Photo by author.*

The magnificent painting by Sherry Kemp of William Davis driving the 1908 LaFrance engine responding to a fire alarm from the Junior engine house. *Courtesy of the artist, Sherry Kemp.*

Junior Fire Company for sixty-seven years. Driver Davis was the first paid firefighter in Frederick County killed in the line of duty.[169]

On April 12, 1913, the Junior Fire Company purchased land on North Market Street from Henry Lampe, Nettie L. Lampe and Francis M. Lampe for $1,000. On May 5, another lot was purchased from Henry Lampe, acting as executor of the will of Elizabeth M. Lampe. These two parcels were combined to provide a suitable location for a new engine house. The property at Second and Market Streets was sold at auction to William O. Kolb for $20,800. Kolb planned to convert the building into the Empire Theatre Company.[170]

Construction began on a new engine house located at 535 North Market Street. The architect of the new firehouse was B. Evard Kepner. The building was constructed by Merhl Gittenger. During the construction, the horses and apparatus were stored at stables near Hagan's Blacksmith Shop near the corner of Fourth and Market Streets. By July 30, 1913, the bell had been placed in the new bell tower atop the fire station.[171] The September 9, 1913 edition of the *News* reported that finishing touches were being made on the new building: "All scaffolding has been removed from the building and the brickwork cleaned." Though the news report indicated the building may be ready for occupancy in a few weeks, "it is understood to be the intention of the company to take its new apparatus into the building, and no horses. This may delay the occupancy of the building to about November 1."[172]

On September 13, 1913, as a result of the death of driver William Davis, the Junior Steam Fire Engine Company No. 2 purchased a Christie motorized tractor to mount to the front of the American LaFrance steamer and eliminate horses in the new fire station.[173] At least five horses, Dan, Billy, Frank, Prince and John, were used to pull the steamers and hose wagons from 1903 to 1913. The horses were sold and used other than fire service.[174]

In September 1913, two of the horses were used to pull the new 2,400-pound two-wheeled street sweeper, originally driven by one of the Juniors' paid drivers.[175] On October 18, 1913, all five horses were sold at auction. The pair of gray horses were bought by Charles Klipp for $146 apiece. The old gray horse was sold for $116 to Charles Fagan. William H. Shipley purchased the brown horse for $200.[176] One of the gray horses, Dan, was ultimately purchased by a vendor in Washington, D.C. While he was working for his new owner, Washington, D.C. Fire Department engines whisked by, and Dan instinctively wanted to respond to the fire. In an attempt to free himself from his harness, poor Dan, always the fire horse, fell and broke his leg, ultimately ending his life. The members of the Junior Fire

Left: "New" Junior fire house at 535 North Market Street shortly after being built in 1913. *Collection of author.*

Below: Junior 1908 LaFrance steam pumper attached to the 1913 Christie tractor. *Courtesy Frederick County Fire and Rescue Museum.*

Company were very distraught upon hearing of Dan's death.[177] Another positive event that occurred in 1913 was the rebuild of the 1877 Charles T. Holloway hook and ladder, configured to allow the ladder wagon to be hooked to and propelled by the motorized apparatus.[178]

On Thursday, November 27, 1913, the American LaFrance engine with the new Christie tractor was shipped from the Front Drive Tractor Company of Hoboken, New Jersey, on the Pennsylvania Railroad. It was probably not a coincidence that the agent for the Pennsylvania Railroad, Abram Kemp Keefer, was a member and future president of the Junior Fire Company No. 2. Delivery in Frederick was anticipated on Monday, December 1.[179] The new fire engine house was formally opened in December 1913 with a grand "gala."[180] Though designed to accommodate horses, the new fire engine house of the Junior Fire Company never saw horses in the building. The White chassis placed under the hose and chemical wagon and the Christie tractor under the LaFrance Metropolitan steamer replaced horses with motorized horsepower. The membership of the Juniors succeeded in ensuring the tragedy of Uncle Billy Davis would never occur in the new firehouse.[181]

A story of one of the last runs of the Christie engine, handed down to retired paid driver of the Junior Fire Company No. 2 Sergeant James Bopst, tells of a fire at the North Market Street School that was located almost directly across from the Junior firehouse. On Sunday, April 27, 1924, around eight o'clock in the morning, James Murphy noticed smoke coming from the rear school building. He ran across the street and reported the fire to the Junior firemen on duty at the engine house. The fire was initially attacked with "chemicals and plug stream," which indicates the White chemical wagon and a hose line hooked directly to the fire hydrant were initially used.[182] As the story was told by Sergeant Bopst, to enable the Christie to build enough pressure in the boiler to pump water, instead of going directly to the fire plug just across from the Junior fire hall, the driver made a left-hand turn and drove up North Market Street and around the water fountain at Seventh Street (then a circular roadway around the fountain) and back down Market Street to ensure the steamer was ready to pump water from the hydrant to fight the fire.[183] *In memory of Sergeant Jim Bopst, who passed away just prior to the publication of this book.*

On November 18, 1913, the newly powered White hose and chemical wagon arrived in Frederick from Baltimore after making the trip in about three hours driving under its own power. The LaFrance hose and chemical wagon was re-enameled and painted in red with gold and black trim with "Junior S.F.E. Co. No. 2" on each side. The newly refurbished chemical

Junior Fire Station at 535 North Market Street with Christie engine. *Collection of author.*

The 150th anniversary of the Junior Fire Company No. 2 artwork of 1949 Ahrens-Fox, 1931 Ahrens-Fox and 1924 Ahrens-Fox engines. Junior was believed to be the only fire company to have all three types of Ahrens-Fox engines in service at one time. *Collection of author.*

wagon also received three large searchlights, with the middle light of twenty-four candlepower.[184]

The Christie/LaFrance Steamer served the company well into the age of motorized fire apparatus. The Christie remained in front-line service until the purchase of the 1924 Ahrens-Fox piston pumper for a cost of $13,700. The new engine arrived at the Pennsylvania Railroad yard on East Street on October 23, 1924. The pumper was an open-cab piston with a one-thousand-gallons-per-minute capacity and sixty-five gallons of water in the booster tank. The engine was equipped with a deluge set of three Siamese connections.[185]

The engine was tested by pumping six streams of water at one time from Carroll Creek on the swinging bridge on Bentz Street. The firefighters proceeded to take the engine to Church Street, where the 1924 Ahrens-Fox threw two streams of water over the Town Clock of Trinity Chapel. This is believed to be the last time a fire engine threw water over the Town Clock.[186] This engine was the first of three Ahrens-Fox pumpers purchased by the Juniors.[187] The Christie was sent back to American LaFrance for repairs and remained in second-line service for several more years.[188]

Chapter 8

UNITED STEAM FIRE ENGINE COMPANY NO. 3

"LILY OF THE SWAMP" TO THE WATEROUS

After the Civil War, the United Fire Company most likely operated three hand-drawn pumpers: the Merrick and Agnew engine purchased in 1850, the Old Lady previously owned by the Mechanical Fire Company No. 1 of Baltimore City and another engine possibly from the Mechanical. Frederick was a growing community, and firefighting technology was changing. Steam pumpers were quickly replacing hand pumpers in major cities. By 1876, the Junior Fire Company had purchased the first steamer in Frederick.

The Uniteds formed a committee to investigate the purchase of a steam pumper as early as 1876.[189] The Clapp and Jones Manufacturing Company of Hudson, New York, had one of the premier steam fire pumpers on display at the 1876 Centennial Exposition in Philadelphia. As the exposition provided fire, police and medical services, it is possible Clapp and Jones provided the engine to be utilized in case of fire as well as have it on display. Mertilew R. Clapp initially worked for the Silsby Manufacturing Company, the builders of the Juniors' steam engine. Clapp started to build his own line of steam fire engines, forming a partnership with Edward D. Jones. Between 1862 and 1892, Clapp and Jones built over five hundred steam fire engines.[190]

Officially called the International Exhibition of Arts, Manufactures and Products of the Soil and Mine, the Centennial Exposition was truly a magnificent display to highlight the pride and ingenuity of America. With

At least two hand pumpers of the local fire companies are visible in this photo of the 1871 groundbreaking ceremony of the Maryland School of the Deaf. *Courtesy Frederick County Fire and Rescue Museum.*

over 250 pavilions and seven miles of avenues and walkways occupying 285 acres, agriculture and manufacturing were highlighted with arts and science. The "Centennial" was modeled after the 1851 Crystal Palace Exhibition in London, with ornate and artistic design.[191] Jacob Engelbrecht visited the Centennial Exposition between October 16 and 23, 1876. He advised, "I really do not think I saw the thousandth part. People from all nations of the world were there nearly every day. There were from one hundred thousand to one hundred & fifty thousand people on the ground. Too much to be seen."[192] No doubt, the visitors from Frederick were overwhelmed with this massive display.

The committee from the Uniteds saw the Clapp and Jones engine and was immediately impressed with the beauty and machinery of the engine. The committee agreed the company should purchase the engine and reported their findings back to the membership with such a recommendation. The company agreed with the committee report and purchased the engine, with contingencies. The engine on display in Philadelphia had a beautiful boiler that added considerable expense. As the company only had a little over $3,200 committed to purchase the engine, the manufacturer agreed to make the boiler copper with a nickel/brass plating. The Uniteds also purchased a "spider" hose reel to respond with the steamer.[193]

The steam piston pumper was delivered by rail at the Baltimore and Ohio Railroad depot on February 22, 1878. The Uniteds paid a total of $4,000 for the engine and five hundred feet of gum hose.[194] Upon arrival in Frederick, the excited members removed the tarp placed over the engine during shipment and were immediately distressed. Instead of the beautiful shiny steam fire engine they saw in Philadelphia, an ugly big black machine was revealed. The firemen attempted to remove the covering by scraping but were concerned they would damage the metal.

Disheartened, the United members recovered the engine and secretly removed the engine from the railroad flat car to the fire station across the street from the railroad depot. The fire company summoned the manufacturer for guidance. Clapp personally came to Frederick to investigate. Upon his arrival, he instructed the United members to pour boiling hot water over the engine to remove the black coating that had been placed over it to protect the engine during delivery. The hot water removed the cosmoline protective coating, and as a cocoon would reveal a beautiful butterfly, the magnificent steam fire engine that was so attractive to the United committee

Postcard of the Lily of the Swamp, 1878 Clapp and Jones steamer of the United Steam Fire Engine Company No. 3. *Courtesy of Sherry Kemp.*

Some of the original committee members who purchased the Lily of the Swamp pose with the famous engine in front of the United's firehouse. *Collection of author.*

in Philadelphia appeared. Not unlike firefighters of today, the "Swampers" retired to a nearby tavern to celebrate.

However, the purchase was not complete until the engine was tested for efficiency. Though the engine was a beautiful piece of machinery, the fire company needed to be sure the pumper would provide satisfactory service to provide advanced fire protection to the town. Once the formal testing of the pumper was complete, Clapp bragged the engine would easily throw water across the top of the Town Clock. The engine performed admirably, throwing two strong streams of water over the top of the "old German Reformed Church."[195]

Most likely as a result of this test of the new engine, appropriately named the Lily of the Swamp, on March 4, 1878, the Consistory of the German Reformed Church (today's Evangelical Reformed Church United Church of Christ) voted to prohibit the further testing of fire engines throwing water over the Town Clock. More detailed information of the testing of fire engines over the Town Clock will be provided in a later chapter.[196]

The name "Lily of the Swamp" was indicative of the location of the firehouse built in the swampy area south of Carroll Creek, the nickname "Swampers" for the company members and the "Swamp Hall" name by the

Local residents and a town police officer in front of the United Fire Company; note the number "3" outlined on the firehouse in flowers. *Collection of author.*

locals to describe the firehouse.[197] Upon receipt of the Lily of the Swamp in 1878, the United Fire Company changed the company name to the United Steam Fire Engine Company.[198] At some point, the number 3 was also added to the official company name.

In 1890, the Lily was transported by train to help fight a fire that destroyed the American Hotel and five other buildings in Point of Rocks. It was rumored the Lily pumped so much water it "lowered the level of the Potomac by four inches." The engine also helped save the surrounding buildings of the United firehouse in fighting a fire on the corner of Saint and Market Streets at the United States Hotel. The Lily was an extremely reliable pumper, pumping for 405 straight hours to help supply water in the drought of 1893.[199]

The Lily of the Swamp was in service for a total of fifty-four years—quite an accomplishment for a steam pumper. In 1932, the Lily was donated to the Smithsonian Institute, where it was on display for many years at the American History Museum. To give the cherished engine a proper send-off, the members of the Uniteds gave the Lily her final run. Though for many years she was pulled by horses—Doc, Dud, Jody and Swamp—the "Final Run of the Lily" was pulled by manpower in the form of Swampers. At 1:30 p.m. on February 22, 1932, the "alarm of fire" was sounded, summoning the Lily to fight one last fire.[200]

The engine responded up Market Street to Patrick Street along following a hose reel borrowed from Walkersville. The engine and crew stopped at Carroll Creek to pump water to fight a fire at the "building" that was constructed for this special occasion. The engine pulled up to the creek as crowds watched and waited. After about ten minutes of work by Harry Marman and Thomas Chew, the gauge on the steam engine reached fifty-five pounds of pressure and the steam whistle blew. The fifty-seven-year-old engine performed admirably. The fire was completely extinguished and the engine was back in quarters by 1:55 p.m. The entire demonstration took less than a half hour.[201]

Swampers who "ran with the engine" pulling the Lily on her final run were William Bartgis Storm, Alton Y. Bennett, "Skeets" Hoffman, Johnny Walsh, Dr. James Marrone, Stewart Brown, Walter Lipps, William S. Gosnell, Harry Marman, Charles Chew, Julius Lochner, Charles Lipps, Carl Brown, George Miller, Marion Moberly, William D. McCleery, Joe Walling, Walter Duvall, Lee McCardell, C. Edward Sinn, John Young, Fritchie Hauer and W.L.O. Fisher. William E. Hahn was injured as he fell pulling the engine, breaking his arm. He was taken to Frederick City Hospital.[202]

Last run of the Lily before the engine was donated to the Smithsonian Institute. *Courtesy Frederick County Fire and Rescue Museum.*

The Lily of the Swamp became a legendary engine in Maryland. The design of the diorama for the Volunteer Firefighters Room previously located at the University of Maryland included a model of the Lily as the featured steam-era engine. The diorama that was dedicated in the early 1970s is now located at the Frederick County Fire and Rescue Museum in Emmitsburg.[203] The actual Lily of the Swamp can be publicly viewed at the Fire Museum of Maryland in Lutherville.[204]

In addition to the Lily, the Uniteds purchased additional equipment prior to the first motorized apparatus. The first indication of specific ladder accruement was the 1875 purchase of hook and ladder equipment by the Uniteds. Though no specific manufacturer is noted, this early equipment most likely was an assortment of ladders carried on a separate wagon. The Lily, hose reel and ladder equipment were the staple of firefighting apparatus of the Uniteds until 1906.[205]

In 1906, the company purchased a hose and chemical wagon from the LaFrance Company of Elmira, New York. On March 9, 1906, the steam engine and the hose and chemical wagon were tested by setting a large box

United 1906 LaFrance hose and chemical wagon. *Collection of author.*

on fire at the Ox Fibre Brush Works. The hose wagon arrived three minutes after sending an alarm. The engine was able to throw water within seven and a half minutes from the time the alarm was sounded. The mayor and board of aldermen for Frederick witnessed the test.[206]

In 1911, the United Steam Fire Engine Company No. 3 ventured into the motorized automobile type of firefighting apparatus by purchasing a new Waterous auto fire engine from the Waterous Engine Works Company of St. Paul, Minnesota. The apparatus was equipped with a "hose box" that could carry 1,200 feet of hose and a 35-gallon chemical engine located behind the driver seat with 200 feet of ¾-inch chemical hose. The pump was rated at 500 gallons per minute and with two 2½-inch discharges. A similar engine in service in Washington, D.C., pulled suction lift of 12 feet 7 inches and delivered 506 gallons of water for a test of one hour.[207]

On February 22, 1912, C.H. Waterous arrived to conduct the acceptance testing of the new engine. The City of Frederick provided $4,500 and Frederick County provided $500 toward the purchase of the engine that had a total cost of $8,500. The Uniteds would also sell the American LaFrance hose wagon and horses to offset the cost of the purchase.[208] The engine underwent drive testing on February 24, 1912, with Waterous driving the engine throughout Frederick County. From Buckeystown to Walkersville, the engine was a spectacular sight, drawing crowds to see the new automobile-

The 1912 Waterous engine of the United Fire Company. *Collection of author.*

type fire engine. After returning to Frederick, the engine pumped from a hydrant. Leaves and tadpoles came out of the hydrant, and when the sleeve was removed, a dead muskrat had been sucked and killed by the pressure.[209]

Thomas Meushaw, superintendent of machinery of the Baltimore Fire Department, supervised the six-hour pump test of the engine. The test was conducted off the South Market Street bridge over Carroll Creek.[210] However, during testing, several gears were torn. The testing was halted and underwriters' approval delayed. Waterous was handling the engine when the malfunction occurred.[211]

On April 22, 1912, the engine was being driven by the fire company driver when the differential gears were stripped and some major parts of the machine were broken. The four-hundred-pound shaft was cut cleanly in two. The shaft was shipped to the factory, and new parts were ordered. The engine was pulled by a "large motor truck" to the Ideal Garage, where it would be kept until repairs were made. It was noted this was not an imperfection in the machinery but rather a mistake on the part of the driver.[212] The Waterous engine seemed to have constant difficulties, as another notation on August 14, 1912, indicated a tooth in the pump gear was broken and had to be repaired.[213]

The Waterous automobile chassis barely served ten years, as the April 6, 1923 *News* reported the Uniteds were ordering a new Dodge-Graham

United members pose with the 1923 Dodge chassis with the 1912 Waterous body at the United firehouse. Note number 3 now permanent above the door. *Collection of author.*

Brothers chassis to replace the Waterous chassis as a motorized chemical and hose wagon without a pump. With three chemical and hose wagons in service in Frederick, it was reported at least eight out of ten fires could be handled by these units.[214] Later, the Uniteds added large floodlights to the unit to provide lighting at night fires and emergencies.[215]

In 1919, the Uniteds purchased the first of what would become six different Ahrens-Fox fire apparatus in service in Frederick. The chain-driven piston

The 1919 Aherns-Fox engine in the station. *Collection of author.*

The 1919 Aherns-Fox and the 1923 Dodge/Waterous with four unnamed United members. *Collection of author.*

pumper was delivered on September 4, 1919. The engine was shipped to Frederick on the Pennsylvania Railroad. While the train was being situated on the siding along East Street, an automobile struck the rail car carrying the new engine and pushed the rail car off the tracks. The fire engine was not damaged. It should be noted that until about 1970, the Pennsylvania Railroad operated down the middle of East Street and the railroad station was located alongside the city street. Several sidings crisscrossed East Street and East Patrick Street.

The engine had a front-mounted 750-gallon-per-minute piston pump and was capable of carrying 1,200 feet of hose. The engine was also equipped with a 60-gallon chemical tank and "fully equipped with lanterns, axes and small ladders and presents a fine appearance." The engine was painted "Swamp red" and was slightly longer than the Waterous engine.[216] The engine passed underwriters' testing and was accepted by the board of aldermen of the city of Frederick on September 12, 1919. The aldermen passed a motion to accept the engine and pay the Ahrens-Fox Company of Cincinnati, Ohio, $11,500. In addition to passing the underwriters' test, it was also noted the engine threw a stream of water over the Town Clock using a 1 3/8-inch nozzle.[217] This engine survives today as "Gladhill No. 1," owned by the Gladhill family and on display at the Gladhill Tractor Company near Jefferson, Maryland.[218]

Chapter 9

INDEPENDENT HOSE COMPANY NO. 1

STEAMER TO AUTOMOBILE FIRE ENGINE

When the Junior Fire Company purchased the first steam pumper in Frederick, the 1876 Silsby, the LaFrance Fire Engine Company of Elmira, New York, also had a steam pumper in Frederick for exhibition and to sell. Both pumpers arrived in Frederick on September 20, 1876, and were tested and witnessed by the fire companies in Frederick.[219] The Independent Hose Company No. 1 was also interested in purchasing a steam pumper and participated in the testing of both engines. On September 30, 1876, the chairman of the Independent's committee to investigate a purchase of a steam pumper recommended to the fire company to purchase a steam engine. On January 21, 1877, the fire company voted to purchase a steam pumper from the LaFrance Fire Engine Company.[220] The steam pumper Romeo was purchased by the Independent Hose Company for $4,000. The engine was delivered in early April 1877 and "tested at Birely's meadow at the depot."[221]

On November 23, 1880, an alarm of fire sounded, sending the fire companies to the new gas works on East Patrick Street. The companies responded promptly, but Romeo, the new steamer of the Independent Hose Company, was nearly destroyed.[222] The alarm was received for the Frederick City Petroleum Gas Works at approximately 10:30 a.m. When crossing the bridge on Water Street over Carroll Creek, the engine veered to the left and broke through the railing, crashing into Carroll Creek. Lawrence Bentz was directing the tongue of the engine and was thrown into the creek and pinned under part of the engine.[223] The Independent Hose Company did not own horses. The steam engine and hose reels were normally pulled by manpower. If necessary for a distant response, the Independents would borrow or rent horses from a nearby livery stable.[224]

Doctors were called to render aid. Bentz had a crushed leg, cracked ribs and several abrasions. Romeo was severely damaged. It took nearly an hour and a half to bring the fire under control. Damage to the gas works was estimated between $500 and $1,000. Gas service to Frederick was not interrupted. The members of the Independents met with representatives from LaFrance the following day. The cost of repairs to the engine was so high that the company decided to investigate the purchase of a new engine. On December 11, 1880, the LaFrance Fire Engine Company offered to provide a new engine for $2,000 and the remains of Romeo.[225]

The replacement engine from the LaFrance Fire Engine Company arrived in Frederick on March 18, 1881. It was "one of the handsomest engines in the city. It is a fourth-class crane neck."[226] The engine arrived at the Pennsylvania Railroad Station along East Street. The first pumping test was conducted on the South Market Street bridge over Carroll Creek and threw a stream of water 248 feet through an inch and a quarter nozzle.[227] After several tests, the Independent Hose Company accepted the engine on March 30, 1881.[228]

The engine was named Juliet and was considered a third-class rotary pump engine. The steam engine performed admirably at a fire of the stables at the fairgrounds, pumping water through 2,800 feet of hose to fight the fire.[229] However, one of the most famous stories of Juliet was for a distant response to the town of Emmitsburg to fight a fire at St. Joseph's College and save buildings of the campus that we know today as the National Fire Academy.

A telegraph was received by Colonel L. Victor Baughman of the Junior Fire Company in Frederick of a major fire at St. Joseph's College in Emmitsburg.[230] Captain James McSherry of the Independent Hose Company of Frederick was notified of the request for assistance and mustered equipment from the Independents, Juniors and United Fire Companies to respond. The Independent Hose Company dispatched the LaFrance steam pumper Juliet, the Juniors sent a compliment of long ladders and the Uniteds sent a hose reel. Each company also sent personnel.[231]

The equipment was loaded on freight cars of the Pennsylvania Railroad on East Street, where an awaiting train traveled a reported seventeen miles in eighteen minutes to the Bruceville interchange with the Western Maryland Railway. A locomotive of the Western Maryland pulled the "fire train" to

Members of the Independent Hose Company No. 1 pose with the 1880 LaFrance steamer Juliet. *Collection of author.*

the junction of the Emmitsburg Railroad near Rocky Ridge and proceeded into Emmitsburg and directly on to the campus of St. Joseph's College.

Upon arrival, there was a reported effort to prohibit the Independent engine from securing water from the hydrant. Isaac Annan, president of the Emmitsburg Water Company, "was sitting aside of the hydrant, refused the Independents the right of water. He proclaimed to the people that he was the president of the Emmitsburg Water Company and that not a drop of water should be used without his consent. After a consultation between Captain McSherry, Colonel Baughman, and the director of St. Joseph's, the president reluctantly gave his permission to the fire Company and from that time on until 1 a.m. the Frederick firemen worked like troopers. At the time of their arrival at St. Joseph's the hose Company of Emmitsburg was doing grand work, but their efforts were of course unequal to the requirements. The Frederick boys arrived home at 3:30 a.m. having subdued the flames. All their apparatus was brought with them, with the exception of a quantity of leather hose belonging to the Independence which cannot be reeled."[232]

This fire was not the only time the steamer Juliet was loaded on a freight car and transported to help fight a massive fire. The fire destroyed the Monocacy Valley Creamery, Creagerstown Hotel, Valley Hotel, Ogle's

Store, several houses, barns and outbuildings. It stopped just shy of igniting the Lutheran church.[233] This fire will be described in detail in a later chapter.

Juliet may have also responded to the Great Baltimore Fire of 1904, but several records indicate the contrary. An article in the *News* on July 6, 1926, about the sesquicentennial celebration of the birth of the nation spoke of the fire companies in the line of parade march. The *News* article stated the steamer Juliet saw eight days of continuous service at the fire in Baltimore.[234] However, this information is suspect, as other newspaper articles published during the time of the fire and an operational report of the fire do not indicate any response from Frederick to the 1904 Great Baltimore Fire.

On February 7, 1904, a message was telegraphed to Baltimore fire chief George Horton from the Junior Fire Company of Frederick: "Accept our sincere sympathy. Can we be of any service with our department? It is all at your disposal and command." No reply was ever received. However, the following day, Mayor McLean responded to Mayor Smith of Frederick, "Fire Department not needed. Now under control. Many thanks."[235]

Another document that questions the accuracy of the report that Juliet was at the 1904 Great Baltimore Fire is the "Fire Department Operations Technical Report—The Great Baltimore Fire—February 7, 1904," by Gary Frederick, retired assistant chief of the Baltimore Fire Department. The report lists every piece of apparatus and manpower utilized at the fire. There are no notations or listings of any fire equipment or personnel responding from Frederick to assist Baltimore.[236]

Juliet remained in service until September 1926. Though not responding on initial fire response, the engine was held at the engine house as a reserve unit until it was sold for just sixty dollars.[237] Steam pumpers did not carry a supply of hose for firefighting; thus, fire companies maintained hose reels. The Independent Hose Company had two hose reels prior to purchasing the ornate Romeo hose carriage in 1893. The early hose reels were sold to the newly formed New Windsor Fire and Hose Company in Carroll County in 1908.[238]

In October 1893, the Independent Hose Company purchased a beautiful hose carriage from the Charles T. Holloway Company in Baltimore. The four-wheel Spider reel was mounted on a carriage chassis and was much larger than the two previous hose reels. The unit was built both for responding to fire alarms as well as more ornate accruements for parade and dress occasions. The hose carriage was placed in service in 1894 and was named Romeo in honor of the first steam fire engine purchased by the

A hose reel originally owned by the Independent Hose Company and sold to the New Windsor Fire Company on display at the Frederick County Fire and Rescue Museum. *Photo by author.*

Independents. The fire company could now respond to a fire with Romeo and Juliet! A more in-depth discussion of this hose carriage and Charles T. Holloway will be referenced in a later chapter.[239]

In April 1907, a committee of the Independent Hose Company made a report to the membership to purchase an automobile combination fire engine. The contract was entered with the Howe Engine Company of Indianapolis, Indiana, on May 24, 1907, for the cost of $4,662 and additional improvements at $162. This Howe engine was the first automobile type of fire engine to be placed in service in the state of Maryland.[240] The engine arrived in Frederick by a Baltimore and Ohio freight train at 6:30 a.m. on September 2, 1908.[241]

The engine was a combined hose wagon and pumping engine with a rear-mounted tri-plex pump. The pump utilized the same automobile engine to propel the vehicle on the road as well as provide energy to the pump. The engine could throw three streams of water at pressure from 125 to 150 pounds per square inch (PSI) and was capable of pumping between 450 and 550 gallons of water per minute (GPM). The engine had an overhead ladder rack that carried two sixteen-foot ladders that could be utilized as a twenty-eight-foot extension ladder, one fourteen-foot ladder

Members of the Independent Hose Company No. 1 ready to pull hose carriage Romeo in a 1958 parade. *Courtesy Frederick County Fire and Rescue Museum.*

Independent 1907 Howe engine being tested at the water fountain at Market and Seventh Streets. *Collection of author.*

A 1907 Howe engine responding up North Market Street near the Junior firehouse. *Courtesy Frederick County Fire and Rescue Museum.*

with scaling hooks and one scaling ladder.[242] The engine was tested on September 7, 8 and 9, 1908, with the fire company approving the engine performance.[243] One such test was conducted by pumping water from the water fountain located at Seventh and Market Streets.[244]

On October 6, 1908, members of the fire company drove the new Howe engine for a trial run down the Gas House Road. The engine got stuck and had to be pulled back to Frederick by a horse.[245] The engine was placed in service on December 1, 1908. "Chaffeurs" William Fleischman and James E.S. Brust were deemed "fully capable of manipulating the apparatus."[246] One of the first fire responses for the engine was for a large bank barn on fire on the Thomas farm on O'possumtown Pike, two and a half miles north of Frederick. The Howe initially responded but returned after only traveling part of the way to the scene, as it was realized there was no water in close proximity to the fire and the engine did not carry tank water. In turning the corner from Church Street to Market Street, the engine struck a buggy owned by Thomas C. Yeakle and broke the buggy shafts. The fire company advised the accident did not occur due to "reckless running" but rather due to the crowded condition of the street.[247] No doubt, the response of the automobile-type engine was an event to behold.

At some point after receiving the automobile engine, the Independent Hose Company started charging a fee for fire responses outside the city of Frederick. A charge of $60, equivalent to approximately $900 in 2021 monies, was made for a county response. Though it is not noted if the monies were paid by the property owner or Frederick County commissioners, a discussion ensued between officials of the city, county and Independent Hose Company regarding fees and reimbursement for damages incurred for responses.[248]

Chapter 10

HOOK AND LADDERS TO THE CITIZENS TRUCK COMPANY

Early firefighting apparatus was initially designed strictly to pump water to attempt to extinguish a fire. It was soon realized ladders were needed to provide an avenue of rescue for occupants of multistory houses. The fire service adapted by initially putting ladders on wagons to respond to fires in addition to the pumping engines. This was most likely the arrangement utilized by the United Fire Company that has listed as "Hook and Ladder" equipment in 1875.[249]

The first apparatus built specifically as a hook and ladder was the 1877 Holloway Hook and Ladder built for the Junior Fire Company by Charles T. Holloway, famed Baltimore apparatus builder. This unit was rebuilt with stronger ladders in 1913 and remained in service until the organization of the Citizens Truck Company in 1926. This hook and ladder was in service at both the engine house at Second and Market Streets and the present location at 535 North Market Street.[250]

A notation in the Engelbrecht diary also indicates a "hook and ladder" company was organized in Frederick in August 1879. The president of the company was Frank B. Carlin. It was noted the company "had not received the machine this year." There is no further reference to the name of the company or other members.[251]

On July 3, 1890, a hook and ladder truck was received by the Guardian Hose Company of Mechanicstown. The truck was furnished by J.M. Kerrigan of Emmitsburg, and William H. Loy of Thurmont made the iron framing. The beams of the hook and ladder were lettered "Mechanicstown, Guardian Hose Company No. 1."[252]

In 1897, the first organized fire protection unit was formed in Brunswick. The equipment included a Howe hand pumper, a hook and ladder wagon and hose reels. Horses owned by the B.P. Crampton Company were used to pull the equipment to fires outside the town.[253] On July 13, 1925, the Brunswick Volunteer Fire Company and the Independent Hose Company responded to a large house fire in Burkittsville. News reports noted the Brunswick Volunteer Fire Company responded with a hook and ladder truck "but had no pumping facilities. The Brunswick firemen lent their hose to the Independents."[254]

As early as July 2, 1904, there was a stated need for an aerial hook and ladder. Recommendations made by the board of underwriters advised, "There should be one modern constructed hook-and-ladder truck obtained with city service extension and ordinary length ladders and necessary equipment and driver and team of horses and suggest its location at station No. 1." It also noted horses should be provided to pull all steamers and the present hook and ladder to all fires. The ladder, then located at the Junior Steam Fire Engine Company No. 2, was hand drawn.[255] In addition, the steamer of the Independent Hose Company was either hand pulled or horses were borrowed or rented from a nearby stable.[256]

Among the additional recommendations made to the mayor and aldermen was the "necessity" of installing a modern-constructed fire alarm system with at least fifteen street alarm boxes.[257] The discussion was not the first time the city government was requested to install a fire alarm system. On March 23, 1895, the three fire companies held a meeting to discuss the need to install a citywide fire alarm system. The board of aldermen agreed to get appropriations to fund the system.[258] However, a fire alarm system with street alarm boxes was never installed in Frederick. Frederick is one of the few major cities in the state of Maryland and the only city in western Maryland that did not install a fire alarm system.[259] It is probable the City of Frederick never purchased fire alarm boxes because police call boxes were installed at many locations around Frederick that would be utilized by local citizens.

Another major recommendation of the board of underwriters' report was a consolidated fire department and the need for a "chief engineer," what today would be considered a fire chief. The United Fire Company was opposed to the idea of unification and voted against the other fire companies to discuss the idea.[260] For many years, the idea of a "city" fire chief was brought forth but never acted on. As early as 1851, the Independent Hose Company sent a resolution to the United Fire Company proposing a united citywide fire department, which was unanimously rejected by the Swampers on April 5, 1851.[261] The suggestion of a fire chief was similarly rejected several times.

The ranking officer in the fire company was initially known as foreman and later captain. The term "chief" was not even used in the volunteer companies of Frederick until the 1970s. There has never been a formal City of Frederick Fire Department, though apparatus and some helmets did have "FFD" or "Frederick Fire Department" printed.[262] The City of Frederick finally hired Walter Murray as the fire/ambulance coordinator in 1980.[263] A city fire chief or city fire department never came into existence.

In 1911, the members of the Independent Hose Company discussed the need for a ladder truck at a regular company meeting. It was noted the need was critical, especially in the business district of the city. It was reported that a Baltimore company had a ladder truck for sale for a "remarkably low price." The truck included ladders and an eighty-five-foot aerial and was "modern in every detail." The company was in no position to purchase the ladder truck at that time; however, it felt it was important to alert the citizens of this need and lobby for the purchase of a ladder truck in the future.[264]

On September 6, 1915, the *Daily News* again made mention of a need of a fire chief for Frederick, noting there was "nothing more vital than

Early helmets of the Junior Fire Company; note "FFD" and rank of foreman and assistant foreman. *On display at Junior Fire Company No. 2, photo by author.*

organization and system." The same article made reference to a recent blaze in Frederick that demonstrated that "handicap the local firemen labor in not having adequate ladders for reaching a building above the second story." The article made a plea for a modern ladder truck as "one of the greatest needs of the Frederick Department."[265]

The initial meeting of the citizens of Frederick for the purpose of organizing a ladder truck company was held on March 15, 1926.[266] On April 30, a public announcement was made for a meeting of "the new fire truck company" to be held at the YMCA at 8:00 p.m. with the intent of forming a permanent organization. By this meeting, the company had more than three hundred charter members.[267]

The Citizens Ladder Truck Company of Frederick was officially organized on Friday, April 30, 1926, at a general meeting held at the YMCA. The elected officers of the newly organized fire company were Dudley Page, president; vice-presidents Walter Zeigler and H. Dorsey Etchison; secretary Richard P. Ross; and William D. Zimmerman, treasurer. The board of directors of the company consisted of the elected officers, two members of each engine company, the mayor and two aldermen. This initial board of directors would serve for the first year.[268] The company officially incorporated in 1931 and adopted the official name of Citizens Truck Company of Frederick City, Frederick County, Maryland.[269]

On May 2, 1926, H. Dorsey Etchison and Dudley Page spoke at the Tuesday meeting of the Kiwanis Club held at the Francis Scott Key Hotel in Frederick. It was noted that over four hundred buildings in Frederick needed the protection of a ladder truck and over two thousand people slept in a building on a third floor or higher.[270] An editorial in the *Daily News* on June 26, 1926, supported the newly formed ladder truck company and encouraged the mayor and aldermen to consider the purchase of "a motor driven hook and ladder truck."[271]

At a meeting of the mayor and board of aldermen on Wednesday, October 6, 1926, a petition was submitted to the city government signed by approximately six hundred residents that included names of members of the new ladder truck company. Petitions supporting purchase of a ladder truck were also received from the chamber of commerce, the Lions Club, the Rotary Club and the Kiwanis Club, all of which unanimously supported the purchase of a ladder truck.[272]

The efforts to purchase a ladder truck finally gained enough momentum for the board of aldermen to pass a motion to purchase a "new fire-ladder truck." Shortly thereafter, the mayor, four aldermen, the presidents of each of the city engine companies and the president of the new truck company left for a trip to inspect ladder trucks in Washington, Philadelphia, Allentown, New York, Atlantic City and Ocean City. The trip was provided by manufacturers of ladder trucks.[273]

On April 21, 1927, bids for the new ladder truck were opened, with Mack-International Motor Truck Corporation the lowest bidder at $15,200. Ahrens-Fox Engine Company was the highest at $16,500. Seagrave Company bid $15,750, and the American LaFrance Engine Company bid $15,500. Action on the bids was delayed until Saturday, April 28, 1927, at 3:00 p.m.[274] Mayor Culler advised the city would only appropriate $15,000 toward a new ladder truck. Following a long executive session of the mayor and aldermen, a motion was passed to purchase the Ahrens-Fox ladder truck with the agreement that the members of the Citizens Ladder and Truck Company would make up the $1,500 difference. The motion passed on a three to two vote. Mayor Culler made it clear the fire company could not solicit the additional funds from the community and the difference must be made up by the membership. President Dudley Page advised this provision was acceptable and that he already had offers from members for volunteer contributions.

The article in the *Daily News* on April 25 revealed more details of the meeting to purchase the ladder truck. After the trip was made to several cities, members of the ad hoc committee voted for the Ahrens-Fox ladder truck. This was fully backed by the membership of the Citizens Ladder and Truck Company, whose members offered to make up the difference in the price between bidders. Alderman Shipley had made the motion to purchase the Ahrens-Fox for $15,000 with the understanding the Citizens membership would make up the difference. Alderman Renn seconded the motion. Aldermen Renn, Shipley and Mullinix voted in favor of the purchase, with Aldermen Munshower and Butcher opposed. Specific details were not released, as they were considered private. It was noted that President Keefer of the Junior Fire Company advised the Juniors would provide housing for the new truck.[275]

At a special meeting of the Independent Hose Company on August 31, 1927, a resolution was adopted pledging support for the "new Citizens Ladder and Truck Company and welcome it to our midst as a new volunteer firefighting organization in our fast growing city." The resolution also "further resolved. That we take this occasion to publicly thank the Mayor and Aldermen for their generosity shown to our company when it has been in financial need."[276] Though the engine companies in Frederick adopted

Left: Driver's hat of Sergeant Charles "Mike" Hahn of the Citizens Truck Company. Note the "A" in the Maltese cross. *Display at Frederick County Fire and Rescue Museum, photo by author.*

Below: A 1927 Ahrens-Fox hook and ladder aerial truck and officers of the Citizens Truck Company; note "Ladder 1" on the aerial. *Collection of Frederick County Fire and Rescue Museum.*

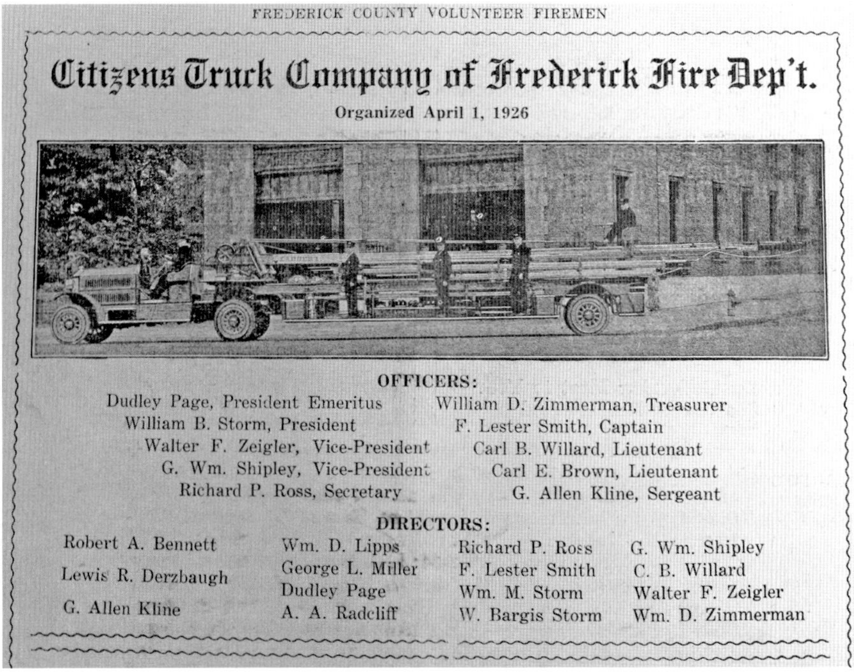

FREDERICK COUNTY VOLUNTEER FIREMEN

Citizens Truck Company of Frederick Fire Dep't.

Organized April 1, 1926

OFFICERS:

Dudley Page, President Emeritus
William B. Storm, President
Walter F. Zeigler, Vice-President
G. Wm. Shipley, Vice-President
Richard P. Ross, Secretary
William D. Zimmerman, Treasurer
F. Lester Smith, Captain
Carl B. Willard, Lieutenant
Carl E. Brown, Lieutenant
G. Allen Kline, Sergeant

DIRECTORS:

Robert A. Bennett
Lewis R. Derzbaugh
G. Allen Kline
Wm. D. Lipps
George L. Miller
Dudley Page
A. A. Radcliff
Richard P. Ross
F. Lester Smith
Wm. M. Storm
W. Bargis Storm
G. Wm. Shipley
C. B. Willard
Walter F. Zeigler
Wm. D. Zimmerman

numbers, the Citizens Truck Company initially utilized the letter "A."[277] Numbering engine companies and using the alpha designation for ladder companies was common in area fire departments such as Washington, D.C., and York, Pennsylvania, though early pictures of the Ahrens-Fox show "Ladder 1" on the aerial.

The long-anticipated Ahrens-Fox aerial ladder arrived in Frederick on November 8, 1927. The tractor section arrived in the morning over the Pennsylvania Railroad, and the trailer and ladders arrived by train later in the day. The truck carried over two hundred feet of service ladders (ground ladders) in addition to the aerial. The Citizens Truck Company gave the

The quarters of the Citizens Truck Company is a historic structure built in 1939 as a project of the Works Progress Administration of President Franklin Delano Roosevelt. *Collection of author.*

new truck a test run on Thanksgiving Day, Thursday, November 24, 1927. The ladder was raised at the City Hospital, Visitation Convent, Houck Apartments and Maryland School for the Deaf. At the Montevue County Home, the aerial ladder reached the top of the building and the forty-foot service ladder went to the third story.

The official test of the new ladder truck was held on Wednesday, November 14, 1927. Charles E. Wood and Walter A. Scott of the Ahrens-Fox Fire Engine Co. were in charge of the testing. The test included raising and lowering the aerial ladder at city hall and other large buildings in the city. The chiefs of the Baltimore and Washington fire departments were invited to attend. Hagerstown firefighter Frank Bentz, formerly of Frederick, would inspect the truck.[278]

A ladder pipe was purchased for an additional $350 and was placed into position. The ladder pipe was planned to be tested in Baker Park with the engines of the Juniors and Independents. It was noted the ladder pipe was capable of throwing 1,500 gallons of water per minute. A life net that was purchased for $150 would soon be installed on the ladder truck.[279]

Captain Henry Miller of the Baltimore Fire Department was assigned by Baltimore Chief Emrich to come to Frederick to instruct the members of

Members of the Citizens Truck Company jumping into the life net of the 1927 Ahrens-Fox ladder truck. *Collection of author.*

A 1927 Aherns-Fox ladder truck in the "new" quarters of the Citizens Truck Company. *Collection of author.*

the Citizens Ladder and Truck Company. Captain Miller was scheduled to arrive in Frederick on Thursday, December 8, 1927, to begin instructing the Frederick firemen.[280]

The beginnings of the first fire response areas in Frederick were reported in the *Daily News* on December 15, 1927. A plan was outlined stating that the fire engine nearest the scene of the fire would respond. Earlier in the month, it was discussed that the new ladder truck would only respond on initial alarms on certain noted structures. The plan also referenced second and third alarms. It was noted that restricting companies to certain zones would break down the rivalry between companies. In the same article, it was noted the Independents received sixty dollars for each fire responded to in the county.[281]

Chapter 11

"WATER OVER THE TOWN CLOCK"

Up from the meadows, rich with corn, clear in the cool September morn, the clustered spires of Frederick stand, green-walled by the hills of Maryland." Thus begins the famous poem "Barbara Fritchie" by John Greenleaf Whittier that made the church spires of Frederick internationally famous.[282] From 1839 to 1924, the volunteer fire companies of Frederick routinely tested their new fire pumping engines by throwing water over the German Reformed Church, one of the famed spires on West Church Street that is home to the Town Clock. This iconic steeple is the oldest of the famed "Clustered Spires of Frederick." Today, the building and steeple are owned by the Evangelical Reformed United Church of Christ and located beside the former firehouse of the Independent Hose Company. For many years, the Trinity Chapel, as it is known today, was the highest structure in Frederick. It is only appropriate that the first and last documented engines to throw water over the steeple were owned by the Junior Fire Company.

Trinity Chapel was built in 1763 by the German Reformed Church. The structure was sixty by forty-five feet with a steeple of stone and completed at a cost of $3,324. In 1778, two bells were purchased from London and placed in the steeple. Several years later, these bells were replaced with new bells from New Haven, Connecticut. Francis Scott Key was baptized in this church in 1779. The original Town Clock was paid for by the citizens of Frederick in the 1790s. It was built by Frederick Heisley and installed in the top of the original steeple of the German Reformed Church.

Trinity Chapel and Town Clock and the former location of the Independent Hose Company No. 1 on West Church Street. *Courtesy of Scott Grove.*

The present steeple of Trinity Chapel, designed and constructed by local architect and builder Stephen Steiner, was completed on March 7, 1807. The original Town Clock was reinstalled into the new steeple by Heisley and local clockmaker John Fessler.[283] The Maryland state legislature enacted an approval on January 12, 1806, that authorized funds for repair of the Town Clock. These funds most likely were utilized to make alterations and repairs to the clock prior to installation in the new spire.[284]

No documentation has been discovered to indicate either the Grandfather engine that was purchased in 1764 or the fire engine purchased by the Sun Fire Company around 1805 was ever tested by throwing water over the Town Clock. However, in 1839, the newly organized Junior Fire Company received a new engine from the John Rodgers Company of Baltimore. The engine arrived in Frederick on Saturday, August 24, 1839. On August 26, the new engine was tested when it threw water over the steeple of the Evangelical Reformed Church steeple. The engine threw water "a little higher than the extreme knob, which was at least 150 ft."[285]

This action most likely concerned the early members of the German Reformed Church, as the congregation had the steeple painted by John Fryer in 1834. Engelbrecht noted that Fryer "swings himself around with pleasure. He gets down & climbs up the ropes in real sailor style." The painter worked from April 28 to May 14, 1834. As he completed his work, he measured the steeple height at 150 feet.[286]

Just over a week after the painting was completed, lightning struck the steeple and "shattered the top of the roof considerable." The lightning struck on the west side of the tower and exited on the east side about midway down the steeple.[287] Though there is no mention of the lightning repairs, it is noted the steeple received a slate roof in October 1835.[288] After the testing of the Junior pumper in 1839, throwing water over the Town Clock became a regular occurrence with new fire engines.

The next documented testing of a pumper over the steeple was on June 14, 1851. Engelbrecht noted, "The 'Juniors' tried their engine this afternoon, and it threw water several feet higher than the German Reformed Chapel steeple—I saw it done—no mistake." It is not known if this was a new engine or a test of the Rodgers engine purchased in 1839.[289]

Several more engines were purchased by the Independents, Juniors and United Fire Companies; however, no specific documentation of testing any other hand pumping engines is available. On May 2, 1853, a John Rodgers engine owned by the First Hagerstown Hose Company was on visitation to the Junior Fire Company and threw water 120 feet—not quite over the roof of the steeple.[290]

In 1861, the steeple "leaning precariously toward the south," as noted in "Timeline of Trinity Chapel," would likely have been due to the pumping action that would take place from the engine location on Church Street flowing water southward over, and probably on, the steeple. The cost of the repairs was $281.91 for the steeple and $460.91 for the clock. The town contributed $100 toward the cost. The remaining costs were paid for in full by elder Abraham Kemp, who presented the amount as a free will offering to the church. The church honored him by naming the new church building on the corner of Market and Church Streets Kemp Hall.[291] It was in this building that the Maryland General Assembly met in a special session in the summer of 1861 and voted against secession from the Union.[292] Following the vote, a suspicious fire seriously damaged the Frederick Courthouse a block away.[293]

The next advanced technology for fire engines was the evolution of the steam pumpers. These powerful engines were soon purchased by each of the three engine companies in Frederick. There is no record of the Silsby engine purchased by the Juniors in 1876 using the church steeple to test the engine capabilities. However, the next two steam pumpers purchased by Independents and Uniteds contributed to a church action to cease "throwing water over the town clock."

The Independent Hose Company purchased a steam pumper from the LaFrance Fire Engine Company. The engine, named Romeo, was thoroughly tested by pumping at P.H. Pyler's coal yard on April 10, 1877. The next day, members of the Independents pulled the engine out of the firehouse on Church Street and threw water more than twenty feet over the Evangelical Reformed Church chapel located just a few doors below the firehouse.[294]

On February 22, 1878, the new steam pumper of the United Fire Company arrived by rail. The engine had been on display at the 1876 Centennial Expo in Philadelphia when purchased by the Uniteds. The fire company members were initially concerned about a black covering that could not be removed from the engine boiler. Mr. Clapp, of the Clapp and Jones Manufacturing Company, builders of the engine, arrived in Frederick and instructed the members how to remove the protective covering.

To demonstrate the capabilities of the new engine, named the Lily of the Swamp, the engine was taken to West Church Street to follow the tradition of throwing water over the Town Clock. Not only did the Lily succeed, but this massive engine threw two streams of water over the Town Clock with ease.[295]

Testing of fire engines again took a toll on the steeple. Shortly after the test of the Lily of the Swamp, the church consistory assessed the costly damage to the clock and steeple as a result of the most recent testing. As a result of this damage, on March 4, 1878, the consistory took action to prohibit such testing of fire engines in the future.

In 1882, a new two-story building was erected with the original stone tower and 1807 steeple remaining. The church, costing $11,000, was dedicated in May 1882 and named Trinity Reformed Chapel. In 1893, a set of ten bells known as the Trinity Chimes were made by McShane Company of Baltimore. The bells ranged in weight from two hundred to two thousand pounds. In the German tradition, the bells have tolled every Sunday since 1893.[296]

Whether the action by the church consistory was forgotten or simply ignored, motorized fire engines were also tested by throwing water over the Town Clock. The United Steam Fire Engine Company No. 3 accepted

Above: This photo shows damage to the steeple of the Trinity Chapel from testing of fire engines throwing water over the Town Clock. *Collection of author.*

Opposite: The 1924 Ahrens-Fox and 1931 Ahrens-Fox engines at the Junior engine house. *Author's collection.*

delivery of a 1919 Ahrens-Fox engine in September 1919. Though the actual underwriters' pump testing was conducted on the South Market Street Bridge over Carroll Creek, the engine also threw two streams of water over the Town Clock.[297]

In 1924, the Junior Fire Company received delivery of an Ahrens-Fox piston pumper. The underwriters' test was conducted from the "swing bridge" over Carroll Creek on North Bentz Street. The engine exceeded capacity, pumping over one thousand gallons of water per minute. The company members then took the engine to West Church Street and directed two streams of water over the steeple.[298] This pumping feat is believed to be the last time any engine threw water over the Town Clock.

Chapter 12

BIRTHPLACE OF THE MARYLAND STATE FIREMEN'S ASSOCIATION

The year 1893 was a very eventful one in the history of the Frederick fire service. It marked the seventy-fifth anniversary of the Independent Hose Company, and the Maryland State Firemen's Association was organized in Frederick. The two events actually comingled into one large fire service celebration.[299]

At a meeting of the Independent Hose Company on July 7, 1892, J. Roger McSherry offered a resolution "declaring the necessity for better organization among volunteer fire companies in Maryland, to increase their efficiency, to insure fraternal intercourse and provide necessary support, and stating that the only manner in which to accomplish this is by the formation of a State Firemen's Association, as now exists in many other states." The resolution requested the appointment of a committee of five from the Independent Hose Company to confer with the Junior and United Companies of Frederick, Guardian Hose Company of Mechanicstown, Vigilant Hose Company of Emmitsburg, Liberty Fire Company and Middletown Fire Company to take steps toward formation of the association and arrange the first meeting at the seventy-fifth anniversary celebration in June 1893.[300]

The idea of a statewide firemen's organization was first made public on July 24, 1892, in an announcement in the *Daily News* about the celebratory events planned in June 1893. The two-day celebration anticipated ninety fire companies and a "greater number of uniformed men in any other firemen's parade held outside of the State of New York." This was the first firemen's tournament event ever held in Frederick, with thousands of

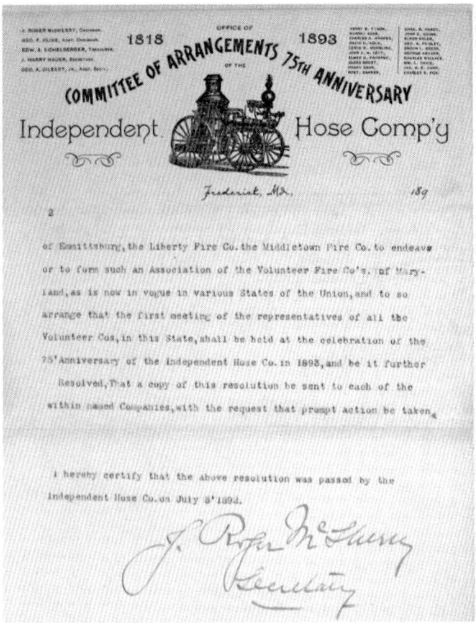

Office of Committee of Arrangements 1818 1893 75th Anniversary
Independent Hose Comp'y
Frederick, Md. 189

2

of Emmittsburg, the Liberty Fire Co. the Middletown Fire Co. to endeavor to form such an Association of the Volunteer Fire Co's. of Maryland, as is now in vogue in various States of the Union, and to so arrange that the first meeting of the representatives of all the Volunteer Cos, in this State, shall be held at the celebration of the 75'Anniversary of the Independent Hose Co. in 1893, and be it further

Resolved, That a copy of this resolution be sent to each of the within named Companies, with the request that prompt action be taken.

I hereby certify that the above resolution was passed by the Independent Hose Co. on July 8'1892.

J. Roger McSherry
Secretary

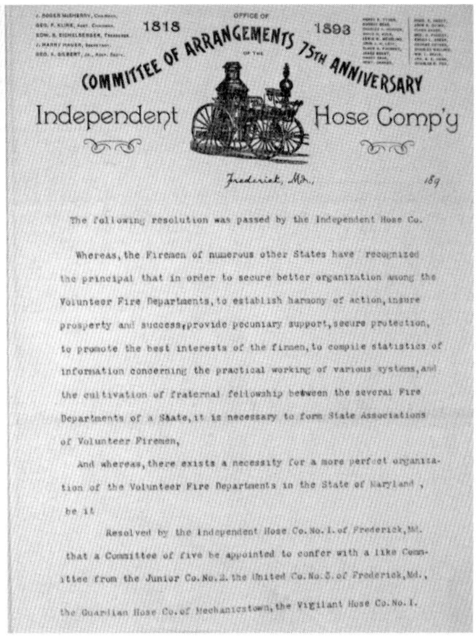

Office of Committee of Arrangements 1818 1893 75th Anniversary
Independent Hose Comp'y
Frederick, Md. 189

The following resolution was passed by the Independent Hose Co.

Whereas, the Firemen of numerous other States have recognized the principal that in order to secure better organization among the Volunteer Fire Departments, to establish harmony of action, insure prosperty and success, provide pecuniary support, secure protection, to promote the best interests of the firmen, to compile statistics of information concerning the practical working of various systems, and the cultivation of fraternal fellowship between the several Fire Departments of a State, it is necessary to form State Associations of Volunteer Firemen,

And whereas, there exists a necessity for a more perfect organization of the Volunteer Fire Departments in the State of Maryland, be it

Resolved by the Independent Hose Co. No. 1. of Frederick, Md. that a Committee of five be appointed to confer with a like Committee from the Junior Co. No. 2. the United Co. No. 3. of Frederick, Md., the Guardian Hose Co. of Mechanicstown, the Vigilant Hose Co. No. 1.

The original letter sent in 1892 to the fire companies in Frederick County asking them to help organize the Maryland State Firemen's Association. *Courtesy United Fire Company, photo by author.*

visiting firemen expected. "One of the prominent features of the occasion and the one that will result in much practical good to the volunteer firemen throughout the State, will be the initial steps towards the formation of a State Firemen's Association."[301]

The committee organizing the seventy-fifth anniversary sent public prints to firemen throughout the state calling attention to the proposed statewide organization. The intent of such an organization was to promote the interest of the volunteer fire service in the state. The editorial in the *Daily News* advised they "deserve support and encouragement of the people in a liberal measure."[302]

On March 22, 1893, it was reported that J. Roger McSherry of the Independent Hose Company would serve as the chief marshal for the massive firemen's parade to be held on June 7, 1893, to celebrate the seventy-fifth anniversary of the Independent Hose Company.[303]

On March 24, members of the Junior Fire Company, Independent Hose Company, United Fire Company and Guardian Hose Company of Mechanicstown attended a meeting held at the Junior Hall to hear the report of a subcommittee on the proposed constitution and bylaws to organize a statewide firemen's association.[304] An earlier announcement of the proposed statewide firemen's association noted the Vigilant Hose

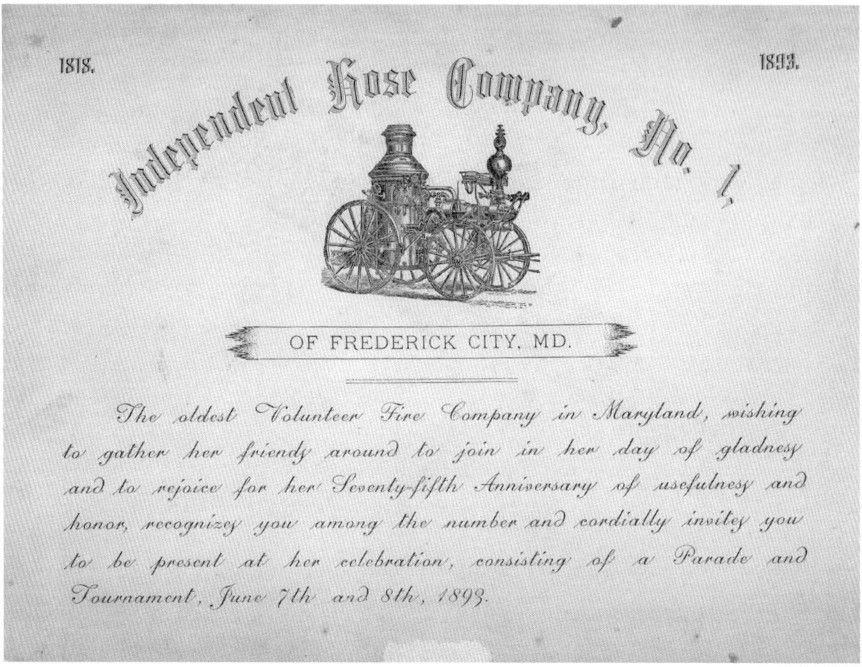
1818. 1893.

Independent Hose Company, No. 1,

OF FREDERICK CITY, MD.

The oldest Volunteer Fire Company in Maryland, wishing to gather her friends around to join in her day of gladness and to rejoice for her Seventy-fifth Anniversary of usefulness and honor, recognizes you among the number and cordially invites you to be present at her celebration, consisting of a Parade and Tournament, June 7th and 8th, 1893.

The 1893 invitation to the seventy-fifth anniversary celebration of the Independent Hose Company No. 1. *Collection of author.*

Company of Emmitsburg was also involved in the early discussions.[305] In a related item, the Junior Fire Company No. 2 of Reading, Pennsylvania, was to arrive on March 25 to secure local accommodations to participate in the parade and tournaments.[306]

On March 29, 1893, another meeting was held at the Junior Hall to hear a report of the committee to organize a statewide firemen's association and take action on the proposed constitution and bylaws of the Maryland State Firemen's Association. The meeting was called to order by the committee chairman, J. Roger McSherry, of the Independent Hose Company of Frederick. Following remarks by Chairman (to later call president) McSherry, correspondence was read from J.Q.A. Holloway of the Veteran's Firemen's Association of Baltimore and from companies in Hagerstown and Westminster stating their support of the proposed state association. This action formally established the Maryland State Firemen's Association as an organization.

In honor of the seventy-fifth anniversary and the grand occasion of the establishment of the Maryland State Firemen's Association, the

Independent Hose Company purchased an ornate hose carriage from the noted Baltimore apparatus builder Charles T. Holloway. Holloway was also elected the senior vice president of the Maryland State Firemen's Association, representing the Baltimore Veteran's Association. The Independents named the new hose carriage Romeo after the fallen first steam fire engine that was destroyed in an accident.

The groundwork for the Maryland State Firemen's Association was laid out prior to the arrival of the volunteer fire companies from throughout Maryland. The object of the new organization "is to revive the volunteer fire companies, and create a more united brotherhood, and to pay benefits to firemen who are hurt while in the discharge of their duties as firemen." A five-dollar initiation fee would be charged to join, and each delegate to the convention was required to pay one dollar each. "It is hoped that the next legislature will make a yearly appropriation, as is done in other States for similar organizations."[307]

The great occasion was finally at hand. On June 7, 1893, the first meeting of the Maryland State Firemen's Association was held in the City Opera House in Frederick. The initial organizing officers were J. Roger McSherry, president; Edward Koontz and Benjamin H. Blackston,

The 1893 Independent Hose Company hose carriage Romeo on display at the Frederick County Fire and Rescue Museum. *Photo by author.*

vice presidents; William Crimmins, recording secretary; and H.R. Heck, treasurer. The first meeting was called to order at 8:20 a.m. by President McSherry. It should be noted that McSherry addressed himself as president even though this was the first official meeting of the association and official elections for officers had not been held. In fact, the first objection before the assembly came when Mr. Armstrong of Hagerstown stated the attendees from Hagerstown believed they were here to assist in forming the association, not participate in deliberations of an existing association, as the Hagerstown companies had not anticipated paying dues. Mr. Koontz of Frederick moved that since there was some misunderstanding, the initiation fee and dues be waived for a period of time after the convention. The motion passed.[308]

Prior to the report of the Credentials Committee, the convention assembly heard an address by Lawrence W. Clark of the *Firemen's Herald* of New York. He spoke on the advantages of having a statewide firemen's organization, as New York volunteer firemen had been organized for a period of time. He spoke of the purposes and character of such an organization. At the end of the convention, Crimmins made a motion that the *Firemen's Herald* be the official organ of the Maryland State Firemen's Association, which carried.

The Veteran Volunteer Firemen's Association was made up of members of the former volunteer fire companies in Baltimore. The paid fire department in Baltimore took over fire protection on January 2, 1860.[309]

Mr. Worthington of Frederick moved for the adoption of the report, which was approved. Mr. Brooks of Westminster presented a resolution that the association would incorporate under the General Corporate Laws of the State of Maryland, which also passed. The Maryland State Firemen's Association was officially incorporated under the laws of the state of Maryland on March 8, 1894.[310]

Other actions that passed at the first convention included that copies of the proceedings of the convention be printed and five copies be distributed to each member company. Hagerstown was voted as the location of the second annual convention, and the annual meeting of the association would be held the second Thursday in June each year in a place so designated by membership at the previous convention.

General Steiner of Frederick made a motion that all of the presiding officers be maintained for the incoming year. A motion was also made to elect Charles T. Holloway of the Baltimore Veteran Volunteer Firemen's Association as the senior vice president, which passed unanimously. In

Parade of the first Maryland State Firemen's Association convention and seventy-fifth anniversary of the Independent Hose Company No. 1 in June 1893. *Collection of author.*

addition to the vice presidents from Frederick, each member company also provided a vice president to serve. Although it is hard to imagine today, the entire convention took place in two hours and ten minutes, adjourning at 10:30 a.m. to allow delegates time to see the various activities being held.[311]

Fire companies and apparatus in the 1893 parade. *Collection of author.*

The actual convention received only cursory coverage in the Frederick newspapers, but the tournament and parade were grand events that received extensive publicity. One of the highlights of the convention was an exhibition of the hook and ladder company of the Rex Fire Company of York, Pennsylvania. It was noted that the "Rex boys were nimble as cats," climbing up the seventy-five-foot spring-loaded aerial and demonstrating a rescue from a burning building. They also exhibited utilization of the ladder as a water tower, throwing streams of water atop city hall.

The *News* reported on the parade and listed winners in each category. The Baltimore Veterans were victorious over the Washington Veterans on the trial of hand engines, which was held on West Church Street, throwing water 158 feet, 9 inches. The Columbia Fire Company No. 4 of Alexandria, Virginia, won the steam pumping contest, getting upstream and making attachments the quickest, in eight minutes and ten seconds. Several thousand people gathered at the Frederick fairgrounds to witness the hose race won by First Hose of Hagerstown, hook and ladder contest won by Water Witch No. 1 of Annapolis and the prize drilling contest won by Junior No. 2 of Reading, Pennsylvania. Guardian Hose of Mechanicstown came in third place.[312]

Competition reel team of the United Fire Steam Fire Engine Company No. 3. *Collection of author.*

First-place winner Columbia Steam Fire Engine Company No. 4 of Alexandria, Virginia, at the 1893 convention. *Collection of author.*

Above: South Market Street near United Fire Company decorated for the 1903 Maryland State Firemen's Association convention. *Collection of author.*

Opposite: North Market Street near Junior Fire Company No. 2 decorated for the 1903 Maryland State Firemen's Association convention. *Collection of author.*

Though the convention was the first meeting of the Maryland State Firemen's Association, the entire event was incorporated into the seventy-fifth anniversary celebration of the Independent Hose Company No. 1. The first engine in Frederick was purchased in 1764. Grandfather was "resurrected" by the Independent Hose Company years prior to the convention and donated to the Baltimore Volunteer Firemen's Veterans Association. Grandfather was located at the beginning of the parade behind the Independent Hose Company and parade marshals.[313]

Frederick County has continued to provide Maryland State Firemen's Association leadership as, in addition to J. Roger McSherry, several MSFA presidents since 1893 have been from Frederick County. Lloyd T. McGill, Joseph W. Gaver, W. Bartgis Storm and William Moore from the Uniteds all served as president. F. Lester Smith, Richard Yinger, David Keller

and Robert Jacobs served as MSFA president representing the Citizens Truck Company. Philip H. Beard of the Walkersville Volunteer Fire Company was the only MSFA president from a company outside the city of Frederick, although he was also a member of the United Steam Fire Engine Company No. 3.[314]

Chapter 13

EARLY MAJOR FIRES

The first recorded fire in the diary of Jacob Engelbrecht was on January 26, 1821, at the dwelling of Dr. Jacob Baer. Though there were at least two fire companies in existence at the time, the entry in the diary indicates the fire was extinguished "by the vigilance of the citizens," with no mention of any fire companies responding.[315] Over a year later, on February 12, 1822, the diary notes, "The people cried fire last evening. It proved to be the warehouse of Stuart Gaither. It was only a little rumpus."[316] On June 27, 1823, three apartments and law offices on Court Street owned by Benjamin S. Pigman, Esquire, were destroyed by fire. "Nearly all the furniture was saved but the buildings are nearly demolished." Again, there is no indication of a fire company assisting in the extinguishment.[317]

On April 1, 1825, the officers of the Frederick Water Company were elected, with Colonel John McPherson elected as president. Dr. William Tyler, later the surgeon of the Junior Fire Company, was elected to the board of directors.[318] By June 25, 1825, "the Fredericktowne Water Works" was progressing with Thomas Shriver serving as superintendent and engineer.[319]

On December 25, 1826, a fire at a tannery belonging to David Bucky six miles north of Frederick was so large that it was visible in town.[320] The first record in Engelbrecht's diary of local fire companies actually fighting a fire occurred on April 1, 1830, at a fire at the City Hotel (Talbotts): "By the exertion of fire companies it was soon arrested," with damage to two or three rooms. As fire companies were noted in the ledger, at least one other company in addition to the Independent Hose Company was in existence.[321]

Though the exact date that Frederick had a hydrant system for firefighting is unknown, Engelbrecht noted a fire on January 24, 1831, in an alley off Market Street between Fourth and Fifth Streets at which "the hydrants were nearly all frozen and few pumps in the neighborhood so that nothing was used but snow which was plenty." The house burned to the ground in about thirty minutes.[322]

On July 20, 1832, Engelbrecht's diary notes that George Hauer on Court Street had his dwelling insured against fire by the "Baltimore Fire Insurance Company" for "100 cents in the $100 or 1 per cent $1000 ensured on it." It is unknown if the property Hauer insured was with the Baltimore Equitable Society, Fire Insurance Company of Baltimore or perhaps another insurance company formed in the early 1800s.

Around midnight on March 2, 1837, a large fire consumed several buildings and damaged several houses in Frederick. A fire at the woolen factory of George M. Conradt destroyed several buildings used for weaving and spinning, as well as the contents of seven looms, a spinning machine and other property. The homes of Mrs. Andrew Heims and George C. Gelwicks were also damaged.[323]

The next major fire noted occurred on March 2, 1841, at the Baltimore and Ohio Railroad locomotive house near what today would be South Carroll Street. The building and blacksmith shop were totally destroyed. Two railroad locomotives and one tender stored in the locomotive engine house were also destroyed. The estimated cost of the loss was $6,000 to $8,000 (approximately $250,000 in 2020). It was exactly four years to the day since the last large fire at Conradt's woolen mill.[324]

Just over two weeks later, on March 22, 1841, a major fire struck the stable and lecture room of the German Reformed Church. The fire consumed the stable and extended to the lecture room and steps, affecting the schoolroom and roof. The fire also damaged the stable and kitchen of Lewis Medtart next to the church facility.[325]

On December 15, 1843, the Mutual Insurance Company of Frederick was founded and began protecting properties in Frederick from the peril of fire. The company still exists today as the Frederick Mutual Insurance Company and is the third-oldest insurance company in Maryland and the ninth-oldest insurance company in the United States.[326] After the company began providing insurance coverage, Engelbrecht often noted if a building was insured and the amount of covered loss.

Possibly the first mention of the Uniteds' nickname "Swampers" was by Engelbrecht in referencing a fire at the stable of Dr. William Bradley Tyler.

The 175th anniversary reproduction of the original fire mark of the Frederick Mutual Insurance Company on display at Frederick County Fire and Rescue Museum. *Photo by author.*

"The Swamp Rangers were the first with their hose, which put it out." About three-quarters of the roof and some of the upper loft was burned.[327]

On September 18, 1849, a fire heavily damaged the cabinet shop of Frederick Schroeder located on Market Street between Second and Third Streets. The upper floors and the roof area were severely damaged. "The building was ensured in the 'Mutual Fire Insurance Company' and injury sustained was valued at $400 which the Company paid Mr. Schroeder and several hundred dollars were made up by the citizens for losses sustained by Mr. S. and his journeymen not ensured."[328]

Fire again struck the German Reformed Church on May 9, 1850, when the carriage house belonging to the parsonage of the church caught fire. The interior of the stable was entirely burned, and a barouche and sulky belonging to Reverend Danial Zacharias were destroyed. Though the cows and horses were taken out safely, a small dog died in the fire.[329]

On January 15, 1853, the iron foundry owned by Benjamin G. Fitzhugh and Samuel Snyder was totally destroyed by fire. The property was insured with the Mutual Insurance Company of Frederick for $2,500, which did not cover half of the loss. Approximately fifteen or twenty people were put out of work. However, by June 20, 1853, the company was nearly rebuilt and anticipated casting iron within a week.[330]

Several buildings were destroyed by fire after being struck by lightning during a severe thunderstorm on July 5, 1853. The icehouse located on the extreme south end of Brewer's Alley owned by Robert Johnson was struck by lightning and totally destroyed. About the same time, a barn owned by George Willard near the Manor Church was destroyed. An old tobacco house owned by Henry Nixdorf near Riehl's Mill was struck and burned to the ground. During the same storm, lightning struck a horse owned by Samuel Zimmerman.[331]

On March 17, 1854, the Baltimore and Ohio Railroad viaduct over the Monocacy River caught fire and was totally destroyed, reportedly from sparks from the engine. Work began immediately to construct a small bridge to allow passengers to cross. A crew of about fifty workers under the supervision of

Superintendent Wendel Bollman began the effort to reconstruct the railroad bridge over the Monocacy River. A new trussel bridge was constructed in four days, as cars passed over the bridge on March 21, 1854.[332]

On April 11, 1858, several houses were damaged by fire along Patrick Street. The fire originated in the home of Peter Nichols and spread to another house east of the City Hotel, damaging the upper story of a front and back building. The fire then spread east to an adjoining house owned by John Markell and west to a house owned by Harry Thomas. The Nicholses' house was insured by the Baltimore Fire Insurance Company for $2,000, and the Markell property was insured by the Frederick County Mutual Insurance Company.[333]

The diary of Jacob Engelbrecht highlights numerous other fires in the 1800s. The 1850s noted many fires of suspicious or incendiary origin. Several stables and barns suffered fires and were destroyed. In 1851, the Mutual Insurance Company of Frederick County offered a reward for information leading to the arrest and conviction of individuals who may be responsible. Though three individuals were charged and two convicted on September 4, 1851, the incidents noted in the Engelbrecht diary continued through the decade.[334]

The most tragic fire in Frederick history occurred at the Zeller's Store fire on May 28, 1888, when an estimated seventy firefighters were injured in a fire and explosion. On May 28 at approximately 6:45 p.m., Charles Poole, a young employee of a store owned by Charles Edward Zellers on the northeast corner of Market and South Streets, went to the cellar of the store with a lantern to draw five gallons of gasoline from a barrel and accidentally ignited gasoline vapor that had leaked out of the barrel. The boy immediately left the cellar and summoned the fire department. Smoke was already coming from the cellar and a residence of the Zeller family next to the store.

All three companies responded to the "general alarm" and were just beginning to throw water on the fire when a rumbling sound followed by a terrible explosion occurred. Several barrels of oil, gasoline and whiskey stored in the cellar had exploded. The wall of the warehouse facing South Street collapsed into the street, and the lower front of the building hurled across Market Street into the crowd of onlookers. The violent explosion injured approximately seventy people, including at least thirty-one members of the United Fire Company. Firefighters who were actually in the building at the time of the explosion exited alive but covered with cuts, bruises and broken limbs.

Young Charles Poole died of his injuries. John William Moore Sr. of the United Fire Company was severely burned but survived his injuries. Charles S. Howard, the business manager of the *News*, received injuries of the hand and was disabled from work. Several police officers at the scene were also injured. The *News* issued a special edition and listed all the names of those injured in the incident. Several doctors responded to the scene, and the city drugstores opened up to provide supplies. All three city firehouses opened to assist victims. Though firefighters from all three companies suffered injuries, no firefighter lives were lost.[335] Ironically, on May 28, 1888, the day of the fire, the *News* ran a brief article highlighting the fact that Mr. Zeller had installed railings and netting around the flat roof of the storeroom to protect his children from falling off the roof and keep people from climbing on the storeroom roof.[336]

A major fire swept through the town of Point of Rocks on March 23, 1890. The American Hotel, four dwellings and a general store were destroyed as flames were fed by high winds. Mr. W.T. Mullinex, an agent of the Baltimore and Ohio Railroad in Frederick, received a telegram from the Point of Rocks agent that a fire was raging and assistance was desperately needed. The United Fire Company prepared the engine Lily of the Swamp and apparatus to be transported by train. The United Lily, hose reel and ladders were loaded onto a gondola that was attached to a passenger car, and a special train was soon en route. The request for assistance was received around 12:00 p.m., and the train arrived in Point of Rocks by 1:35 p.m. Within ten minutes, the Uniteds' engine pumped from the Potomac, throwing "two streams of water."[337] It is said the Lily of the Swamp pumped so much water it "lowered the level of the Potomac by four inches."[338]

On May 2, 1902, a major fire struck the town of New Market, destroying the Protestant Episcopal Church, two dwellings and a saddler shop. The fire started in the saddler shop and spread to the house of Minnie Roelkey. The flames spread to the church that was located behind the house, and the church burned to the ground. The efforts to save the rectory beside the church were successful, but the fire spread to a dwelling on the other side of the Roelkey house and destroyed the home of George Smith. Nearly every member of the town participated in the firefighting efforts, with men and women carrying water to pour on the fire.

A call was made to Frederick to send fire engines, but after Mayor Smith and representatives of the Frederick fire companies discussed the situation over the phone with persons on the scene in New Market, it was decided engines would be of no service, as there was no water supply for an engine to operate. The organ and all the furniture in the church were saved. The church had been previously insured, but the policy had lapsed about three years before the fire.[339]

A disastrous fire struck the town of Mount Airy, located on the Frederick-Carroll County line, in February 1903, nearly destroying the business district. Fifteen buildings, including two hotels, were destroyed. The fire started about 2:30 a.m., and by 3:00 a.m., it was determined the fire was likely to sweep through the town if assistance wasn't summoned. A telephone call was made to Frederick, whereupon President Joseph W. Gaver of the United Fire Company organized a special train from the B&O Railroad to transport the Lily of the Swamp and hose carriages to Mount Airy.

Prior to the arrival of the Uniteds, the community fought the fire with bucket brigades. The special train arrived after 6:00 a.m., just in time to save William H. Hood's large general store. Four railroad locomotives, each with a capacity of seven thousand gallons of water, supplied water to the United steamer.[340]

Mount Airy was struck again with a massive fire in 1914 when ten buildings and one dwelling in the business district were destroyed. Flames spread from Farmers' Grain and Milling Company to the lumber and coal yard of E.M. Molesworth. A dwelling occupied by W.W. Baker and his wife and family of six children that was owned by Peoples' Lumber and Supply was destroyed. Also destroyed by the fire were the Runkles' and Wagners' ice plant, Watkins' and Banks' grocery, R.L. Runkles' Hardware Store, Skegs grocery store and the First National Bank, which was a brick building that was totally gutted. A quarter mile away, barns, henhouses and a garage owned by Arnold Fleming were ignited by the flaming embers and destroyed.

Apparatus from Frederick responded to the scene. The Waterous automobile engine of the United Fire Company responded under its own power and was the first on the scene. An engine of the Independent Hose Company was loaded on a flat car and transported on the Baltimore and Ohio Railroad. The engine from the Independents pumped water from tank cars that arrived from the railroad watering station. The railroad kept four engines close to the train station, throwing water on the building.[341]

A third disastrous fire struck the Mount Airy business district on June 4, 1925, damaging nine buildings and causing an estimated $200,000 loss. Marion Runkles's grain mill, Zepp's store, Runkles's hardware store, Riddlemoser's electric shop, Mount Airy Meat and Grocery, Norman Hood's fertilizer office, Gurney Brashear's grocery and the First National Bank were destroyed. Word was sent to Frederick, and President Thomas

Chapline, president of the Independent Hose Company, dispatched the automobile engine and a complement of manpower. Engines also responded from Rockville, Kensington and Ellicott City. Once again, the Baltimore and Ohio Railroad locomotives hauled water for fighting the fire.[342]

The northern Frederick County community of Creagerstown was initially settled in 1750. John Kreiger (original spelling) laid out the town before the Revolutionary War. The town was originally at the junction of the Baltimore and Pittsburgh Roads and the Washington and Buffalo Roads. Buffalo Road still exists along the eastern border of Frederick County and was part of the original road that led to Buffalo, New York.[343]

The fire broke out about 9:30 a.m. on June 2, 1914, and was said to have started in the engine room of the Monocacy Valley Creamery. The first buildings lost in the fire were the Monocacy Valley Creamery, a house belonging to Mary Kreepz, the town hall and a house owned by B.F. Bell. The fire jumped the street and consumed the barn, wagon shed, corncrib and house of George Stevens. Ogle's Store and dwelling and all stock were destroyed, as well as houses owned by Mrs. John Grinder, Isaac Hankey, Barn Freeze, Louis Miller and George Hann. Other houses were also slightly damaged.

The fire also consumed two hotels then located in the town: the Valley Hotel and the Creagerstown Hotel. Both hotels and associated outbuildings were destroyed, with a loss of both buildings deemed $11,000. On June 2, 1914, Chesapeake and Potomac Telephone Company Plant Line chief W.T. Loy received a phone call asking him to send aid to Creagerstown for a major fire threatening to destroy the town. Loy summoned Frederick mayor Fraley, who authorized the steamer Juliet of the Independent Hose Company to be sent.

Superintendent R. Paul Smith of the Hagerstown and Frederick Railway secured a gondola and freight motor to send the steamer to Thurmont. The special train left Frederick at approximately 12:00 p.m. While en route, the citizens of Thurmont built a trestle to unload the steamer and secured four horses to pull the engine to Creagerstown. The special train was delayed at Isanogle's Switch due to a car derailment. Even with this delay, Juliet reached Thurmont in fifty-seven minutes.

A team of four horses awaited the arrival to pull the fire engine to Creagerstown. Upon arrival in Thurmont, it was discussed the fire engine would be of little use as the rudimentary roads from Thurmont to Creagerstown were very poor and it would take considerable time to reach the fire. As the citizens were attempting to unload the engine, word was received that the fire had nearly burned itself out. The train began a return trip to Frederick with the engine at 2:00 p.m.; Juliet was never unloaded from the train.

While awaiting the arrival of the Independent engine, the residents of Creagerstown were instructed to dam up a small stream to provide sufficient operating water for the steamer. Unfortunately, by the time the steamer arrived in Thurmont, it was decided it would be of little use and was never unloaded. The fire burned up to the Lutheran Church before dwindling out. County commissioner Stevens of Creagerstown was in Frederick when he received word of the fire and lost the cover for the top of his automobile rushing to the scene. J. Roger Fisher of the *News* detrained the special train at Isanogle's Switch when the train was stopped by the derailment and was taken by Frank Rice to Creagerstown, where he reported on the fire for the newspaper.

The article in the June 2, 1914 *Frederick News* also noted Lewis R. Dertzbaugh, assistant foreman of the Junior Fire Company No. 2, and Guy Motter, also a member of the Junior Fire Company No. 2, were en route to Emmitsburg and heard of the conflagration and responded with aid. The fire essentially burned itself out with only the community to try to fight the fire and remove property from threatened buildings. The fire burned freely without any fire engines or fire companies ever arriving to assist.[344]

The Myersville Fire of 1919 caused extensive damage in the town of Myersville. On January 4, 1919, the car barn, station and electric substation of the Hagerstown and Frederick Railway were destroyed, as well as the bank and grain elevator, as a massive fire swept through the town. Snow on the ground and roofs combined with the hard work and determination of the citizens using bucket brigades saved many private homes and businesses.

The initial reports from the *Valley Register* in Middletown reported the conflagration in Myersville to residents of the valley. The paper reported nearly $30,000 in total damage. It went into extensive depth as to the cause of the fire, the electric substation of the Hagerstown and Frederick Railway. It was reported the railroad stationmaster, Alvie Horine, reported "arching" throughout the day.[345]

Woods and mountain fires today are not nearly as volatile as in the dry spell of the 1920s. On November 25, 1922, the *Frederick Post* reported on two large fires in the Catoctin Mountains. A fire near Wolfsville consumed over 1,500 acres in about fourteen hours, while a fire near Catoctin Furnace burned around 7,000 acres.[346] Fighting forest fires in Frederick County was serious business, as Harrison Robins found out in 1922. He was summoned by the forest fire warden to assist on a blaze on Sugarloaf Mountain and refused to assist. He was charged with "failing to fight a forest fire" and fined $16.75.[347]

Chapter 14

PARADES AND OTHER EVENTS

Fire companies in parades are a part of the fabric of America. The first mention of a parade is in the minutes of the United Fire Company in 1849 when other fire companies and city officials were invited to participate in the celebration. On May 1, 1859, the Uniteds escorted the Independent Hose Company to the train station for a visit to Hagerstown. On June 7, 1851, a parade was held to dedicate the new cupola and bell of the Junior Hall. The Uniteds turned out seventy-three members in red jackets and hats. The Independents had fifty-six men in the line of march and the Juniors sixty-four.[348]

Visitations between distant fire companies became a major social event for local fire companies. Visits to fire companies in other cities such as Baltimore, Chambersburg, Hagerstown, York and even Wilmington were fairly regular occurrences.

The American Band was organized by the United Fire Company in 1858 and continued until 1893. The Independent Hose Company has a fife and drum corp.[349] A drumstick and wooden flute of this unit are on display at the Frederick County Fire and Rescue Museum in Emmitsburg.[350]

On May 1, 1852, the Uniteds visited Philadelphia as guests of the city's American Hose Company. The Juniors and Independents escorted the Uniteds for their journey. Upon arrival in Philadelphia, the Diligent Engine Company of Philadelphia joined the Uniteds in procession, receiving them to the city. The Uniteds had a stopover in Baltimore, where they were invited to breakfast by the Lafayette Hose Company. The Uniteds also took an engine on the trip, which cost each member five dollars.[351]

The 1889 drum corps of the Independent Hose Company No. 1. *Courtesy Frederick County Fire and Rescue Museum.*

On August 7, 1875, the Union Fire Company of York, Pennsylvania, paid a visit to our firefighters and paraded their steam engine and hose reel throughout the town. They demonstrated their steamer by throwing water over the Reformed steeple and adjourned to a firemen's ball at city hall.[352]

On May 18, 1880, the Independent Hose Company, with a unit of ninety uniformed personnel, made an excursion to Chambersburg, Pennsylvania, to attend a "Parade of Friendship." The Woodsboro Band also accompanied. It was not unusual for fire companies to have bands with the unit at functions. The Independents presented a banner to the Juniors fire company of Chambersburg.[353]

The Uniteds visited Wilmington, Delaware, as the guests of the Water Witch Fire Company and participated in a large parade. The Uniteds returned the favor in October of that year and invited the Water Witch Company to Frederick for a parade, banquet and firemen's ball.[354]

The pride of each volunteer fire company also led to some rivalry that could extend into more confrontation than firefighting. On August 14, 1858, the Independent Hose Company and the Junior Fire Company had a battle royale with bricks, bats and stones. The disturbance was so violent that several firemen, including Lafayette Norris, George Lambert, Luther A. Nickel and others, were injured in the brawl.[355]

An article in the *Daily News* on July 5, 1895, was headlined "Shame! Shame! Shame! Frederick Firemen Turn Hoodlums and Assault Each Other." A fire at the residence of John Diggs on Bentz Street between Sixth and Seventh Streets was destroyed by fire and an adjoining property heavily damaged by water. The paper reported, "The disgraceful part was enacted by the Independent and Junior Companies, who, instead of keeping up their excellent reputation as gallant firemen, assumed the role of a set of reckless, heedless men who should be held responsible for wrecking the home of a poor colored man." The altercation took place after "the Junior pipemen were on top of the house…when the Independent men directed a stream of water to the roof.…In a little while the firemen forgot their duty and directed the streams of water entirely on each other." The Junior crew actually went into the interior of the house, breaking windows, to attack the Independents with a stream of water. The property owner began to yell to stop damaging his property. Crowds watching the fire actually began to root for each side in the battle. The newspaper called this "costly and uncalled for amusement" and "shame upon the good name of Frederick firemen."[356]

Chapter 15

ADDITIONAL FIRE COMPANIES ORGANIZED AFTER 1915

Myersville Volunteer Fire Company

Though the disastrous fire in Myersville in 1919 started discussion of the need for fire protection, a fire company was not organized by the Town of Myersville until 1925. Even then, purchasing equipment and formally organizing did not occur until 1931, when the burgess and council requested the Independent Hose Company to demonstrate a chemical engine for the citizens of the town.

On October 26, 1931, the burgess, with the consent of two council members, agreed to purchase an American LaFrance fire engine. The engine was purchased from Frederick E. Benson of Paterson, New Jersey, for a cost of $390. F. Lowery of Myersville was paid $50 to haul the engine to Myersville. On November 5, 1931, the Myersville Volunteer Fire Company was officially formed.

On November 19, 1931, a 1919 American LaFrance engine arrived in Myersville. The engine was originally kept in John Horine's shed for an annual rental of $7.50. The engine was moved to the old creamery building owned by John "Tucker" Shank, finally being stored at the old school building.

In 1935, the fire company purchased the old Central Trust Company building on the corner of Main Street and Wolfsville Road for $2,150. An addition was added to the rear of the building to house the fire engine. On October 28, 1935 the Myersville Volunteer Fire Company was officially incorporated and a fire siren installed on the steeple of the Lutheran church to alert volunteers.[357]

NEW MIDWAY VOLUNTEER FIRE COMPANY

The first community meeting to discuss organizing a fire company was held on August 4, 1926, at the Renner Motor Company Garage in New Midway. Citizens of the area raised several hundred dollars in the preceding year through sales and festivals toward the eventual purchase of a chemical engine. Local businessmen A.A. Haugh, Jesse I. Renner, S.A. Grim and Samuel Fogle were noted as the primary organizers.[358] On September 13, 1926, the local businessmen again assembled at Renner's Garage in New Midway to formally discuss organizing a fire company and purchase of a fire engine. In 1927, New Midway purchased a new Ford Howe fire engine for a cost of $2,700.[359]

The new engine arrived by the Pennsylvania Railroad on August 26, 1927. The New Midway Community Fire Department unloaded and tested the engine on August 27, 1927, with a large crowd to witness the test. The engine could throw two streams of water over a three-story building and pump 150 gallons of water per minute. The engine would respond from Renner's Garage until an engine house was built. Jesse I. Renner was president and acting foreman of the fire company and advised practice drills would be held one night a week.[360] In 1934, Renner was also elected to the board of county commissioners of Frederick County.[361]

ROCKY RIDGE VOLUNTEER FIRE COMPANY

On May 9, 1949, the first meeting of the Rocky Ridge Volunteer Fire Company was held. At that meeting, it was decided to refund Ray Valentine for materials used at the fire pond. It was also voted the fire company would take the "light company" into the fire company. In April 1950, the fire company decided to have the fire company take over the fire pond as well as the streetlights. The fire company accepted the building of the Rocky Ridge Hall Association that was located on Long's Mill Road just south of the town square.

In 1951, the only fire call for the new fire company was a disastrous fire of the sliding board at Mount Tabor Park. The fire company was on the scene for seven hours. In March 1951, there was discussion to purchase a Model A Ford fire engine from the Vigilant Hose Company of Emmitsburg. Land was purchased from Mr. and Mrs. William Martin, and a building was constructed by Leon Stover and Roy Dinterman to house the engine.

The company purchased the engine and equipped it with two hundred feet of two-and-a-half-inch hose, two hundred feet of one-and-a-half-inch hose, two two-and-a-half combination nozzles and a gated Y for $764.30. Many fundraising events were held, including shows by famous country music stars Ernest Tubb and Hank Snow.[362]

CARROLL MANOR FIRE COMPANY

After a serious fire at the home of Luther and Nellie Willard on Adams Street in Adamstown, Maryland, in 1953, a community meeting was held at the Odd Fellows Hall to discuss the possibility of organizing a fire company. Mel Schwearing and other representatives of the Independent Hose Company No. 1 of Frederick, the closest fire company to Adamstown, were also in attendance. Clark Gibson called the meeting to order, and a motion was made and passed to organize a fire company.

The fire company name was selected to indicate that the area that would be covered by the new fire company would essentially be within the seventeen thousand acres of the initial tract of land of Charles Carroll of Carrollton that was known as Carrollton Manor. The Carroll Manor was the name of the mansion on the estate. The fire company purchased a 1937 Seagrave Detroit or Safety fully enclosed cab engine from the Independent Hose Company for $3,000. Later, a 1953 Diamond T engine was purchased from the Upper Montgomery County Volunteer Fire Company.

On November 30, 1953, the fire company purchased ground adjoining the Grange Hall on Adams Street. It was at the meeting that the formal name of the Carroll Manor Fire Company was adopted. On April 7, 1954, it was announced that the Hudson Supply of Buckeystown would donate materials to build a fire station. The ladies' auxiliary was formed on May 11, 1954. At the meeting held on July 7, 1954, Mr. Lockwood offered the chassis of a 1947 International to mount a one-thousand-gallon water tank. The first ambulance owned by Carroll Manor was a 1938 Oldsmobile Hennessy donated by the Cline Funeral Home.

The charter membership of the Carroll Manor Fire Company included three African American members: Roger Whalen, Buck Lawson and Bill Bowens. The concern was posed at the April meeting that the Frederick County Volunteer Firemen's Association might require all charter members to be white. At the meeting on July 7, President Gibson advised he had been notified all members of the fire company had to be white in order to join the Frederick County Volunteer Firemen's Association. As a result, the three Black members were asked to resign. Austin Renn also resigned the company in protest. In 2004, at the dedication of Quint 14, the three initial African American members were posthumously made charter members.[363]

New Market District Volunteer Fire Department

A major fire occurred in New Market on March 15, 1918, and the "entire New Market was threatened with destruction," damaging three houses and the Methodist church. In response, a public meeting was held in New Market on March 21, 1918, to discuss the need for some form of fire protection.[364] On April 9, a second meeting was held to discuss the purchase of at least one chemical tank. Due to the fact that there was no water supply, the purchase of a pumper did not seem prudent.[365] On April 26, the committee of Frank Maynard and County Commissioner Wood and former county commissioner Hargarth voted to purchase a double tank chemical engine with two forty-five-gallon tanks. The engine would be located in the middle of town near the Stevens property that had been recently destroyed in the fire.[366] An article in the *Frederick Post* discussing "interest on tax bills" on November 9, 1918, shows an expenditure of $200 for the "New Market Fire Department," which was most likely monies paid by the county commissioners for the chemical extinguisher.

It is unknown how long organized fire protection existed in New Market, as in 1954 at a New Market Lions Club meeting, the question arose once again about the town of New Market and its surrounding area needing fire protection. A public meeting was held, and the citizens of the area agreed that fire protection was needed. The New Market District Volunteer Fire Department was formed, being the fifteenth company in Frederick County to organize. There were approximately sixty-three charter members. The first fire company meeting was held on January 15, 1955, at the old New Market Grange Hall, located on what is now Eighth Alley across from Vintage's parking lot. The company members elected Clyde Smith as president and James R. Hahn as fire chief.

The members set out to raise money as soon as they could, holding dinners and raffles and collecting donations. They raised approximately $12,500 for the purchase of a new fire engine. Some of the men completed a basic fire training course. The fire engine, a 1955 Ford five-hundred-gallon pumper, was ordered from the American Fire Apparatus Company in Battle Creek, Michigan, and arrived in New Market in May 1955. On June 11, 1955, the new engine was placed into service after a ceremony and dinner at the New Market School. Governor McKeldin was in attendance for the dedication. The engine was called Engine #1 and later renamed Engine 151 when Central Alarm was established. The first firehouse was located in a garage behind what was Mealey's Restaurant.[367]

WOODSBORO VOLUNTEER FIRE COMPANY

In 1953, a group of citizens of Woodsboro gathered for a meeting in the Woodsboro Hall located over the bank building to decide what to do with $30,000 in the bank account. A motion was made to give the town up to $5,000 to start a water system and $15,000 to purchase the water to put into the system. The motion also noted to use the balance to buy a fire engine. The first fire company carnival was held from June 23 to June 27, 1953, to help raise funds for the newly organized fire company. And thus, the Woodsboro Volunteer Fire Company, Inc. was incorporated on August 1, 1953. L. Cramer Powell was elected the first fire company president.

The first fire engine was a 1955 American LaFrance on a Ford chassis. Until a fire station was built, the fire engine responded from Cornell's Garage on Main Street. Also in 1955, Lawrence A. Dorsey was elected the second fire company president. Under his leadership, Dorsey was the spearhead of an effort to build a fire station. With the help of several local citizens who were willing to sign $1,000 demand notes to help fund the project, a contract was signed to build a fire station and community hall. The fire station was built in 1956 for a cost of $19,645.78.[368] Woodsboro was an early provider of emergency medical services when it placed a 1954 Cadillac ambulance in service in 1963. Though today the Woodsboro ambulance is A169, the radio designation number was initially A160. Until Central Alarm was formed, a 0 was added after the company number indicated the unit was an ambulance. The unit was purchased from the Thurmont American Legion.[369]

Lawrence A. Dorsey also served as president of the Frederick County Volunteer Firemen's Association and was instrumental in advocating for a countywide central alarm system. Lawrence "Lou" Dorsey served as the burgess (mayor) of Woodsboro and served twelve years on the board of county commissioners.[370]

Graceham Volunteer Fire Company

A fire company in Graceham was originally organized on May 28, 1917, with a fifty-gallon Ajax chemical extinguisher and fifty feet of hose for a cost of $275. The fire company was reorganized on January 1, 1935. A 100,000-gallon capacity fire pond that was thirty feet wide, eighty-five feet long and five feet deep costing $1,700 was constructed. This information was listed in an ad placed in the program book of the 1938 Annual Convention of the Frederick County Volunteer Firemen's Association that indicated Graceham was a member of the Frederick County Volunteer Firemen's Association.[371]

In 1960, a dispute arose with regard to the actual formation of a fire company and some funds in the fire company treasury. Two groups in Graceham—the fire company and a newly formed civic association—claimed funds and propriety over the fire pond. These accusations resulted in legal action.[372] The Graceham Volunteer Fire Company officially incorporated in 1960.[373] In 1962, the fire company purchased a used 1948 American LaFrance engine from the Branchville Volunteer Fire Company in Prince George's County. In 1963, a group representing the Graceham Volunteer Fire Company attended a meeting of the board of county commissioners requesting funding. The group advised that it represented the organization that takes care of the fire pond and streetlights and was not associated with the Graceham Volunteer Fire Company, Inc.

In 1963, the Graceham Volunteer Fire Company became a member of the Maryland State Firemen's Association. The group was using company number 17 when it applied unsuccessfully to be readmitted as a member of the Frederick County Firemen's Association on October 16, 1963. At the time, there was a legal dispute with respect to who actually was the Graceham fire company, as there had been a split in the community and two groups claimed the role.[374] In February 1964, the Graceham Volunteer Fire Company, using the company number 17, filed a decree in court to be admitted in the Frederick County Volunteer Firemen's Association, which eventually led to the company being admitted as a member and being assigned company 18 rather than company 17, which was assigned to Libertytown, as it was admitted to the association ahead of Graceham.[375]

The first new engine purchased by Graceham was delivered in 1966. The engine was a Darley front-mounted pumper on a commercial Ford chassis. The engine carried 500 gallons of water with a front-mounted 750-gallon-per-minute W.S. Darley pump.[376] The company also purchased a tanker from the Libertytown Volunteer Fire Department.

Brunswick Volunteer Ambulance and Rescue Company

The Brunswick Volunteer Ambulance and Rescue Company was formed when a phone call was received on January 6, 1958, from a member of the Maryland Wing Headquarters Civil Air Patrol (CAP) that an ambulance was for sale at Middle River Ambulance and Rescue Association. It was a 1946 Cadillac ambulance "in running condition" that was being replaced by the Middle River association.

The Brunswick Squadron of the CAP was contacted because of its interest in starting a ground rescue unit. The matter was discussed at the CAP meeting, and the squadron commander, Captain Bernard Williams, was designated to meet with Mayor James Cummings and the town council on January 10, 1958, to see if the town would help purchase this ambulance. A request for $100 was submitted, and the money was received. A meeting to organize the Brunswick Volunteer Ambulance Company Inc. was held on January 24, at which time it adopted this name.

Due to the condition of the ambulance they purchased, a second 1946 ambulance was found in Martinsburg, West Virginia, and was purchased for $100 to be used to repair the first vehicle. The first ambulance was placed into service on April 28, 1958, and the first trip was for a broken arm on May 6, 1958; the second was on May 8 for a possible heart attack.

On May 20, the fourth trip was to Jefferson, responding to an air disaster when a Maryland Air National Guard trainer collided with a Capital Airlines passenger plane en route from Pittsburgh to Friendship Airport in Baltimore. Eleven passengers on the aircraft died in the accident that could have been much worse, as it was reported approximately thirty passengers deplaned in Pittsburgh. The pilot of the Air National Guard plane was the only survivor in the crash. This was less than thirty days after the formation of the ambulance company.

Initially, the ambulances responded from Care's Service Station and, later, a two-car garage owned by the Grace Episcopal Church and donated for use. In 1960, the ambulances were equipped with portable oxygen. In 1961, land was purchased on West Potomac Street, and on February 26, 1962, the new quarters of the Brunswick Volunteer Ambulance Company at 200 West Potomac Street was dedicated.[377] With the addition of a rescue truck in 1971, the name was later changed to the Brunswick Volunteer Ambulance and Rescue Company.

Wolfsville Volunteer Fire Company

The village of Wolfsville in the Catoctin Mountains of Frederick County was a distant response for fire protection from any direction. The Myersville Volunteer Fire Company was the closest fire department to Wolfsville, with Smithsburg or Thurmont responding to the areas north of town. On April 1, 1968, the board of directors of the Myersville Volunteer Fire Company brought a proposal to the fire company in regard to fire service in Wolfsville. The Wolfsville area needed to provide a heated building to house one fire engine and provide a siren to alert the volunteers. The Diamond T fire engine that was being replaced by a new pumper would be assigned to the Wolfsville substation once the heated building was complete.[378]

The Wolfsville Holding Company was incorporated to serve as the administrative and financial functions of the Wolfsville substation. The building for the new station in Wolfsville was constructed on Wolfsville Road for a cost of $10,033. The first equipment in the Wolfsville station were a 1941 Diamond T pumper that originally served in Myersville and a 1949 Ford tanker with 1,200 gallons of water that was purchased from the Rocky Ridge Volunteer Fire Company. In December 1979, the Wolfsville Volunteer Fire Company was organized and became a separate entity from the Myersville Volunteer Fire Company, and the Wolfsville Holding Company was dissolved.[379]

Lewistown District Volunteer Fire Department

On April 28, 1967, Lewistown businessman and town resident Jim Grimes announced a group of citizens in the Lewistown Election District had been meeting for about a year in an effort to organize a fire company. It was also announced the initial group had already met the requirements of the Frederick County Volunteer Firemen's Association for a new fire company. It was stressed that this new fire department would be organized to serve all the communities in the Lewistown Election District, not just the town of Lewistown.[380]

The first engine purchased by the Lewistown District Volunteer Fire Department, as it was officially named, was a 1939 American LaFrance engine from the Middletown Volunteer Fire Company for $450. The engine responded on a fire near Lewistown before it was officially in service. The company also purchased an open-cab 1961 Peter Pirsch from the Seat Pleasant Fire Department in Prince George's County. This engine was the first diesel fire engine in service in Frederick County. It carried 500 gallons of water with a 750-gallon-per-minute pump.[381]

Urbana Volunteer Fire and Rescue Company

Charters for a fire department were issued with the State of Maryland on December 11, 1973. Two charters issued for each separate service, fire and rescue. The Urbana Volunteer Fire Department, Inc. and the Urbana Volunteer Rescue Squad, Inc. were officially on record. News of the newly formed fire and rescue companies hit the papers, and membership grew tremendously at the January and February meetings.

The Fifth District VFD of Clarksville, Maryland, donated a 1963 Cadillac for the first ambulance. The Kensington VFD donated a second ambulance, a 1969 Chevrolet Suburban. The rescue squad was officially in service and would answer seventy-one alarms by year's end. The long search for a permanent station site came to an end in 1975, when the department purchased 10.764 acres of land from Arthur Keyes for $53,820.

Within a year, and thanks to a loan from community residents and members Roy Myers and Lawrence Wilcom, Urbana Volunteer Fire and Rescue, Inc. was able to purchase a 1972 Ward LaFrance on a Ford commercial chassis. The engine had a 750-gallon-per-minute pump and carried 50 gallons of water. The pumper was bought for $27,000 from Honeywell Corporation in New Jersey, where the engine was stationed on a contract job site with minimal use. The department was accepted as a member of the Frederick County Volunteer Fireman's Association in March, and Engine 231 was placed in service on April 1, 1974.[382]

Walkersville Volunteer Rescue Company

In 1977, a group of dedicated people began to organize with the intent of starting an ambulance service. A meeting was held on May 17, 1977, with over forty people in attendance. Vaughn Zimmerman chaired the planning committee. Through mass mailings in the fall of 1977, the community was solicited for donations and support. The letter indicated that the organization would be independent of the volunteer fire company, but the equipment would be housed within the fire hall, which is now the town hall.

The request for donations and volunteers was heard by the people of the community. The Glade Valley Lions Club donated a 1969 Suburban ambulance (Ambulance 118), and a 1974 Horton 400 ambulance (Ambulance 119) was purchased from Bethesda Chevy Chase for $12,500. On August 18, 1977, the Walkersville Volunteer Ambulance Service, Inc. was officially chartered. By October, there were forty-three members studying to become emergency medical technicians (EMT). In January, the new volunteer ambulance company began providing service to Walkersville and the surrounding community. The ambulances were placed in service on January 18, 1978.

The units initially utilized the numbers that would have been assigned to the Walkersville Fire Company, Ambulance 119 and Ambulance 118. In the first year, volunteers responded to 289 emergency calls. It was recognized that the company was fast outgrowing the space provided by the fire company. The Walkersville Volunteer Fire Company graciously granted a ninety-nine-year lease on the northeast corner of its carnival grounds. The Walkersville Ambulance Service, Inc. held a groundbreaking ceremony on April 1, 1979. Company member Ray Dayhoff agreed to supervise the construction, and the volunteers donated 4,244 hours of their time in the construction of the building.

In 1983, the company placed the first rescue truck in service, a 1983 Ford EVF rescue squad. The company also officially changed the number designated to the station and equipment and became company 24. In 1984, the legal name of the company was changed to the Walkersville Volunteer Rescue Company, Inc.[383]

Thurmont Community Ambulance Service

The Thurmont Community Ambulance Service was organized in 1969. However, the company's roots date back to July 1953. At that time, a group of men saw the need for emergency ambulance service in Thurmont, and the American Legion Ambulance Service was established as a committee of American Legion Post 168. The original vehicle was a red Packard ambulance, which was demolished in a collision shortly after being placed in service. The next unit was a brown 1949 Packard, which served until being replaced by a 1954 Cadillac in 1960. The first new vehicle purchased was a 1963 Miller-Meteor Cadillac. The ambulance initially responded from a garage located under the first floor of the Thurmont American Legion.

In 1969, a second ambulance was placed in service. The 1966 Miller-Meteor Cadillac was housed at the Guardian Hose Company. The increased call volume also was placing a strain on the dwindling number of Legion members able to staff the ambulance. As a result, the decision was made to reorganize as a separate corporation, and the Thurmont Community Ambulance Service, Inc. was born.

In 1970, a two-bay garage facility was completed at the rear of the Guardian Hose Company parking lot. The facility was rented from the Guardian Hose Company and served the needs of the company until call volume continued to increase.

In the mid-1980s, it became evident that the call volume would soon necessitate a third vehicle. This would also require a new building. In 1986, the property of the Thurmont United Methodist Church, consisting of the sanctuary, Sunday school building with a social hall and the parsonage, became available. Additionally, this site was located next to the Guardian Hose Company. The purchase was completed, and future plans were made to construct a new facility.

When the Frederick County Advanced Life Support program expanded to northern Frederick County, a satellite chase car medic unit, Medic 4, was placed in service and housed at Thurmont Community Ambulance Service.[384]

Ambulance Service in Emmitsburg

Similar to the advent of ambulance service in Thurmont, the Emmitsburg Veterans of Foreign War (VFW) purchased and provided staffing for the first ambulance in Emmitsburg. Several members of the VFW were "on call" to respond as necessary. In the 1970s, an additional ambulance was purchased by the American Legion in Emmitsburg. The ambulance responded from the Vigilant Hose Company until the Emmitsburg Ambulance Company was organized and a separate building built at 300

South Seton Avenue. This location is presently home to the Frederick County Fire and Rescuer Museum.[385] The ambulance company built a new building on Creamery Road in 2010. On January 1, 2018, the Emmitsburg Ambulance Company merged with the Vigilant Hose Company, and the ambulance building became an auxiliary building and social hall for the Vigilant Hose Company.[386]

NOTES

Abbreviations of Major Resources

Engelbrecht: information from Jacob Engelbrecht, *The Diary of Jacob Engelbrecht*, vols. 1 and 2, Frederick, MD: Historical Society of Frederick County, Inc., 2001. Specific dates and pages noted.

FNP Archives: newspaperarchive.com, various Frederick newspapers included. Specific newspaper and dates noted in footnote.

MD Archives: Archives of the State of Maryland, aomol.msa.maryland.gov/html/index.html. Specific session laws and dates noted.

Chapter 1

1. Information from personal visit of author to Schifferstadt, Germany.
2. Information from copy of lottery ticket on display at Frederick County Fire and Rescue Museum.
3. Anniversary copy of Independent Hose Company No. 1 200-year history, 2018 Anniversary Banquet (herein noted Independent 200th).
4. George Washington McCreary, *The Ancient and Honorable Mechanical Company of Baltimore: A Historic Sketch*, introduction, xi, www.google.com/books/edition/The_Ancient_and_Honorable_Mechanical_Com/4V4VAAAAYAAJ?hl=en&gbpv=1&dq=subject:%22Ancient+and+Honorable+Mechanical+Company+(Baltimore,+Md.)%22&printsec=frontcove.

5. John Ashbury, "*...And All Our Yesteryears": A Chronicle of Frederick County, Maryland* (Frederick, MD: Diversions Publications, 1997), 443.
6. T.C.J. Williams and Folger McKinsey, *History of Frederick County, Maryland*, vol. 1 (n.p.: L.R. Tittsworth and Co., 1910), books.google.ki/books?id=_LybcITVSE8C&printsec=frontcover#v=onepage&q&f=false.
7. Engelbrecht, 8.
8. FNP Archives, meeting notice, *Frederick Herald*, March 3, 1832.
9. Ibid., "Sun Fire Company, Interesting Read among Old Records, a Bit of Local History," *Daily News*, January 4, 1909.
10. MD Archives, Maryland State Archives, Session Laws of 1805, vol. 0607.
11. FNP Archives, "That Old Fire Engine," *Daily News*, January 5, 1909.
12. Lewis Fagan III, *150th Anniversary of the Junior Fire Co. No. 2, 1838–1988* (Frederick, MD: Junior Fire Company No. 2, n.d.).
13. Ashbury, "*...And All Our Yesteryears*," 283.
14. MD Archives, vol. 600, 162, ch. 174, passed February 27, 1840.
15. *History of the Independent Hose Company No. 1*, 2, printed by the Independent Hose Company No. 1 on the 175th anniversary (herein noted *Independent 175th*).
16. Williams and McKinsey, *History of Frederick County*, 554.
17. Engelbrecht, 667.
18. Ibid., 685.
19. Ibid., 650.
20. Clarence Forrest, *Official History of the Fire Department of the City of Baltimore* (Baltimore, 1898), 90 www.google.com/books/edition/Official_History_of_the_Fire_Department/RhLBWld4_2AC?hl=en&gbpv=1&printsec=frontcover#v=onepage&q&f=false.
21. Ashbury, "*...And All Our Yesteryears*," 21.
22. Engelbrecht, 858.
23. MD Archives, vol. 214, p. 17, passed January 25, 1836, relief extended.
24. Joy Deater and Jim Deater, *History of Junior Fire Company #2 1838–2013* (Frederick, MD: Gateway Printing, 2013), 2.
25. Ibid.
26. *Gettysburg Compiler*, April 3, 1879, one-sentence mention in electronic copy provided from collection of Jim and Joy Deater.
27. Engelbrecht, vol. 1, 608.
28. "A Timeline of Significant Dates and Events in the Rich History of Trinity Chapel on the 200th Anniversary of the Oldest Steeple of the 'Clustered Spires of Frederick,'" folder developed by Reverend Fred Wenner, pastor emeritus of the Evangelical Reformed Church–United Church of Christ, 2007.

29. MD Archives, Session Laws 1842, vol. 594, p. 45, passed February 4, 1843.
30. Engelbrecht, vol. 2, 636.
31. Deater and Deater, *History of Junior Fire Company*, 2.
32. Ibid., 7.
33. Ibid., 11.
34. Engelbrecht, 701.
35. Ibid., 715, 716.
36. Williams and McKinsey, *History of Frederick County*, 557.
37. Ibid.
38. Ibid.
39. *Independent 175th*, 2.
40. Paul Gordon and Rita Gordon, "Men of the Swamp: A History of the Swampers," November 18, 1995, for 150th anniversary of the United Steam Fire Engine No. 3, Frederick, MD (herein referred to as "Swampers 150th").
41. Ibid., 3.
42. Engelbrecht, 858.
43. Gordon and Gordon, "Swampers 150th," 3.
44. Engelbrecht, 664.
45. Ibid., 696.
46. Gordon and Gordon, "Swampers 150th," 4.
47. Ibid., 11.
48. Ibid., 13.
49. Engelbrecht, 1143.

Chapter 2

50. McCreary, *Ancient and Honorable*, 60–61.
51. Ibid., xi.
52. Ibid., 48.
53. Engelbrecht, 870.
54. Ibid., 898.
55. Minutes of United Fire Company meeting, April 1, 1957, retrieved from fire company archives by author.
56. FNP Archives, *News*, May 20 and August 18, 1899.
57. Ibid., *News*, "Fire Equipment Sold by Libertytown," February 9, 1932.
58. Story as recalled by author.
59. McCreary, *Ancient and Honorable*.

60. Notes taken from minutes of the July 1, 1886 and October 4, 1886 meetings of the United Steam Fire Engine Company No. 3 "donating" the gallery engine to the Baltimore Veterans Association, docs.google.com/document/d/1sjiHxjnABL5TYCxZdO3db1aVMHWLxbQ2ysIY1anap4A/edit.
61. Forrest, *Fire Department of the City of Baltimore* (Baltimore, 1898), 90, www.google.com/books/edition/Official_History_of_the_Fire_Department/RhLBWld4_2AC?hl=en&gbpv=1&printsec=frontcover#v=onepage&q&f=false.
62. Ibid., 182.

Chapter 3

63. Williams and McKinsey, *History of Frederick County*, 323.
64. MD Archives, Session Laws, 1802, vol. 559, p. 2.
65. MD Archives, Session Laws, 1824, vol. 629, p. 16.
66. Middletown Volunteer Fire Company 150th anniversary publication, *The Middletown Volunteer Fire Company, 150 Years of Serving the Valley* (Middletown, MD, 2019).
67. Forrest, *Fire Department of the City of Baltimore*, 42.
68. Engelbrecht, 691.
69. Middletown Volunteer Fire Company 150th anniversary.
70. Vigilant Hose Company 125th anniversary book, *Celebrating Our 125th Year* (Emmitsburg, MD, 2009), 3.
71. Clarence "Chip" Jewell, *Firefighting in Frederick County* (Charleston, SC, Arcadia Publishing, 2005), 12.
72. Vigilant Hose Company 125th, 9–10.
73. "Mother Seton's Daughters of Charity and the Civil War," setonspath.tripod.com/civilwar.html.
74. Ibid.
75. Rocky Ridge Volunteer Fire Company, *50th Anniversary, Rocky Ridge Vol. Fire Co. 1949–1999* (Rocky Ridge, MD, 1999), 3.
76. MD Archives, vol. 611, p. 62, 1846, ch. 69, February 16, 1847.
77. FNP Archives, "Disastrous Fire: Two Dwellings and Several Buildings Burned at Liberty," *Frederick Citizen*, May 19, 1899.
78. FNP Archives, "Fire Company for Liberty," *Daily News*, May 20, 1899.
79. Archives of the United Steam Fire Engine Company No. 3 from the archive electronic secure storage.

80. Water Witch Fire Company in Port Deposit, Maryland, www.wwfco.com/history.html.
81. FNP Archives, "Liberty's Fire Engine," *Daily News*, August 18, 1899.
82. Conversation between author and George Collins of African American Firefighters Historical Society. Members of the Libertytown Volunteer Fire Department identified the Black member as Charlie Key.
83. FNP Archives, "Fire Equipment Sold by Libertytown," *News*, February 9, 1932.
84. Picture of the old firehouse circa 1935 is on display at Frederick County Fire and Rescue Museum.
85. Williams and McKinsey, *History of Frederick County*, 332.
86. Author conversation with John Kinnaird, June 15, 2020.
87. Copy of *Thurmont Clarion*, July 10, 1890, in author collection, provided by Mayor Kinnaird.
88. Ibid., March 6, 1890.
89. Jewell, *Firefighting in Frederick County*.
90. Ibid., 20.
91. Maryland State Firemen's Association 125th anniversary book, *Celebrate 125 Years of Service to Maryland*, vol. 2, *The Lean Years, 1919–1944* (Crofton, MD, 2017), 36.
92. *Baltimore Sun* archives, "Fireman Killed, Harry C. Root Struck by a Falling Smokestack in Thurmont—Excelsior Works Burned," June 19, 1898, www.newspapers.com/clip/53711380/root.
93. Edward D. Smith, "Historic Resource Study Brunswick, Maryland, Historical Data, Chesapeake and Ohio Canal National Park MD-DC-WVA," Denver Service Center Historic Preservation Team, United States Department of Interior, Denver, Colorado, January 1978, irma.nps.gov/DataStore/DownloadFile/443079.
94. Brunswick Volunteer Fire Company History, www.bvfd5.org/content/history.
95. Jefferson Volunteer Fire Company, www.jvfc20.com/history.php.
96. Ibid.
97. FNP Archive, "Jefferson Fire Station Burns, Truck Destroyed," *Frederick News*, April 12, 1969.
98. Ibid.
99. FNP Archive, "Fire at Walkersville, Residence of Mr. John Jamison Destroyed This Morning," *Daily News*, March 7, 1908.
100. Walkersville Volunteer Fire Company, www.walkersvillefire.com/content/history.

101. Interview by the author with Vaughn Zimmerman, past president and life member of the Walkersville Volunteer Fire Company, September 30, 2020.
102. Herbert Harwood, *Blue Ridge Trolley* (San Marino, CA: Golden West Books, 1970), 14.
103. FNP Archives, "Building Activity, Many Lots Progressing at Braddock Heights," *Daily News*, November 12, 1910.
104. Braddock Heights Volunteer Fire Company, bhvfc.org/history.

Chapter 4

105. Gordon and Gordon, "Swampers 150th," 11.
106. Deater and Deater, *History of Junior Fire Company*, 15
107. Gordon and Gordon, "Swampers 150th," 11.
108. Biography.com, "John Brown," www.biography.com/activist/john-brown.
109. USHistory.org, page 31d, "The Pottawatomie Creek Massacre," www.ushistory.org/us/31d.asp.
110. National Park Service, "John Brown's Raid, Harper's Ferry Historical Park," www.nps.gov/articles/john-browns-raid.htm.
111. Gordon and Gordon, "Swampers 150th," 11.
112. Williams and McKinsey, *History of Frederick County*, 348.
113. Engelbrecht, 858.
114. Special Collections of the Maryland State Archives, msa.maryland.gov/msa/speccol/sc5500/sc5572/000001/000000/000017/html/t17.html.
115. National Park Service, "John Brown's Raid."
116. Williams and McKinsey, *History of Frederick County*, 346.
117. Deater and Deater, *History of Junior Fire Company*, 14.
118. Gordon and Gordon, "Swampers 150th," 12.
119. Deater and Deater, *History of Junior Fire Company*, 13–14.
120. Gordon and Gordon, "Swampers 150th," 11.
121. Ibid., 12.
122. Engelbrecht, 859–61.

Chapter 5

123. Williams and McKinsey, *History of Frederick County*, 361, 363.
124. Deater and Deater, *History of Junior Fire Company*, 13–15.
125. Ibid., 15.
126. Engelbrecht, 903.
127. Ibid., 918.
128. Robert Lebling, "America's Zouaves," March/April 2017, www.aramcoworld.com/Articles/March-2017/America-s-Zouaves.
129. Deater and Deater, *History of Junior Fire Company*, picture on p. 12, the uniform noted is a re-creation of a Junior Defender uniform on display at the Junior Fire Company No. 2.
130. Engelbrecht, 903, 915.
131. Fagan, *150th Anniversary of Junior Fire Co.*, 9.
132. Engelbrecht, 917.
133. Ibid., 923.
134. Michael G. Williams, "Baltimore Riot of 1861, Bullets vs. Bricks in Baltimore," HISTORYNET.com, www.historynet.com/baltimore-riot-of-1861.htm.
135. Maryland Archives Special Collection, "Arrests of the Maryland Legislature, 1861," msa.maryland.gov/msa/speccol/sc5500/sc5572/000001/000000/000017/html/t17.html.
136. Frederick News-Post.com, "A Timeline of the Frederick County Courthouse," November 29, 2013, www.fredericknewspost.com/a-timeline-of-the-frederick-county-courthouse/article_647d0a64-b13f-5812-8afc-998411459c04.html.
137. FNP Archives, "Today Is 96th Anniversary of Fire on Record St. Which Scared Citizens," *News*, March 31, 1938, 2.
138. Engelbrecht, 898.
139. Ashbury, "*…And All Our Yesteryears,*" 165.
140. Engelbrecht, 899.
141. Ibid., 913.
142. Deater and Deater, *History of Junior Fire Company*, 16.
143. Gordon and Gordon, "Swampers 150th," 4.
144. Terry Reimer, *One Vast Hospital: The Civil War Hospital Sites in Frederick, Maryland after Antietam* (Gaithersburg, MD: National Museum of Civil War Medicine, Signature Book Printing, 2001), 83.
145. Engelbrecht, 954.
146. Reimer, *One Vast Hospital*, 87.

147. Kevin Pawlak, "Brooklyn Firemen in Action in Maryland," September 12, 2020, emergingcivilwar.com/2020/09/12/brooklyn-firemen-in-action-in-maryland.
148. Engelbrecht, 1052,
149. Gordon and Gordon, "Swampers 150th," 4.
150. Ashbury, *"…And All Our Yesteryears,"* 258.
151. Nancy F. Whitmore and Timothy L. Cannon, *Frederick: A Pictorial History* (Norfolk, VA: Donning Company/Publishers, 1981), 63.

Chapter 6

152. Clarence "Chip" Jewell, "Fire Trucks and Trains," *Volunteer Trumpet* (Maryland State Firemen's Association newsletter), October 2014, 5.
153. Deater and Deater, *History of Junior Fire Company*, 17.
154. Engelbrecht, 1158.
155. *Independent 175th*, 28.
156. MD Archives, Session Laws 1878, vol. 399, p. 34, ch. 33.
157. Personal accounts from author Clarence "Chip" Jewell.
158. Deater and Deater, *History of Junior Fire Company*, 234.
159. FNP Archives, "Firemanic Notes," *Daily News*, May 14, 1889.
160. Deater and Deater, *History of Junior Fire Company*, 24.
161. FNP Archives, "Better Protection Against Fires," *Daily News*, April 25, 1903.
162. Deater and Deater, *History of Junior Fire Company*, 24–25.
163. Ibid., 26.
164. Ibid., 31.
165. Clarence "Chip" Jewell and Warren W. Jenkins, *Firefighting in Frederick* (Charleston, SC: Arcadia Publishing, 2004), 38.
166. Deater and Deater, *History of Junior Fire Company*, 27–28.
167. Ibid.; FNP Archives, "Brief Bits," *Daily News,* September 14, 1908.

Chapter 7

168. FNP Archives, "Fell Under Engine, Wm. B. Davis, Driver of Junior Horses Loses His Life, Tragedy at Engine House," *Daily News*, March 21, 1910.
169. Information on display and in archives of the Frederick County Fire and Rescue Museum in Emmitsburg, Maryland.
170. Deater and Deater, *History of Junior Fire Company*, 29.

171. Ibid., 30.
172. FNP Archives, "Roof on New Enginehouse," *Daily News*, June 30, 1913.
173. Ibid., "Juniors Buy a Christy Tractor" and "Much Activity in Building Circles," *Daily News*, September 9, 1913.
174. Deater and Deater, *History of Junior Fire Company*, 25, 31.
175. FNP Archives, "Street Sweeper for Frederick Arrives, Junior Horses to Be Used," *Daily News*, October 17, 1913.
176. Ibid., "Junior Horses Sold," *Daily News*, October 18, 1913.
177. Deater and Deater, *History of Junior Fire Company*, 32.
178. FNP Archives, "Hook and Ladder Truck in Service, Can Be Hooked Up to New Junior Engine," *Frederick Post*, December 18, 1913.
179. Ibid., "New Junior Chemical Engine Will Arrive in Town Tonight," *Frederick Post*, December 8, 1913.
180. Deater and Deater, *History of Junior Fire Company*, 32.
181. Ibid., 30.
182. FNP Archives, "School Building Damaged by Fire; Loss Near $3,000," News, April 27, 1924.
183. Personal story as related to the author by Sergeant Jim Bopst, retired career firefighter who spent his entire career at the Junior Fire Company and was the son of a Frederick career firefighter.
184. FNP Archives, "Junior Wagon Is Home Again…" *Frederick Post*, November 19, 1913.
185. Ibid., "New Junior Engine Soon Put in Service," *Daily News*, October 24, 1924.
186. Ibid., "New Fire Engine Accepted by Juniors," October 30, 1924.
187. Deater and Deater, *History of Junior Fire Company*, 235.
188. FNP Archives, "New Fire Engine Accepted by Juniors."

Chapter 8

189. Gordon and Gordon, "Swampers 150th," 25.
190. Jack C. Norbeck, "Clapp & Jones Steam Fire Engines," January/February 2002, www.farmcollector.com/steam-traction/clapp-jones-steam-fire-engines.
191. Stephanie Grauman Wolf, "Centennial Exhibition (1876)," Encyclopedia of Greater Philadelphia, philadelphiaencyclopedia.org/archive/centennial.
192. Engelbrecht, 1129.
193. Gordon and Gordon, "Swampers 150th," 25.
194. Engelbrecht, 1140.

195. Gordon and Gordon, "Swampers 150th," 25.
196. "Timeline of Trinity Chapel," 5.
197. Gordon and Gordon, "Swampers 150th," 4.
198. Ibid., 21.
199. Ibid.
200. FNP Archives, "'Lily of the Swamp' Makes Last Run; One Fireman Breaks Arm in Fall," *News*, February 22, 1932.
201. Ibid.
202. Ibid.
203. Diorama display at the Frederick County Fire and Rescue Museum, Emmitsburg, Maryland.
204. Jewell, *Firefighting in Frederick County*, 126.
205. Gordon and Gordon, "Swampers 150th," 24.
206. FNP Archives, "Satisfactory Test," *Daily News*, March 9, 1906.
207. Ibid., "Uniteds' Engine Up-to-Date One," *Daily News*, July 1, 1911.
208. Ibid., "May Test Tomorrow, C.H. Waterous of Auto Engine Co. Expected 10:30 AM," *Daily News*, February 21, 1912.
209. Ibid., "25 Miles an Hour Over Rough Roads, New United Auto Engine Holds Up Under Severe Tests," *Daily News*, February 24, 1912.
210. Ibid., "More Tests for Engine, Pumping and Water-Throwing Power Tried Today," *Daily News*, n.d.
211. Ibid., "Tests Are Postponed," *Daily News*, February 29, 1912.
212. Ibid., "Uniteds Engine Disabled," *Daily News*, April 22, 1912.
213. Ibid., "Tooth in Gearing Brakes," *Daily News*, August 14, 1912.
214. Ibid., "Juniors Adopt Alarm System," *Daily News*, April 6, 1923.
215. Jewell and Jenkins, *Firefighting in Frederick*, 62.
216. FNP Archives, "Car with New Engine Knocked from Track," *Daily News*, September 4, 1919.
217. Ibid., "City to Enact New Parking Ordinance…United Engine Accepted," *Daily News*, September 12, 1919.
218. Jewell and Jenkins, *Firefighting in Frederick*, 61.

Chapter 9

219. Engelbrecht, 1128.
220. "History of the Independent Hose Company No. 1," 175th anniversary booklet, 28, Independent Hose Company, Frederick, Maryland, 1994.
221. Engelbrecht, 1133.

222. Ibid., 1153.
223. Ashbury, "…*And All Our Yesteryears*," 404.
224. Information provided to author by the late John Maly, life member and historian of the Independent and member of the Frederick County Fire and Rescue Museum.
225. Ashbury, "…*And All Our Yesteryears*," 404.
226. Engelbrecht, 1157.
227. Ashbury, "…*And All Our Yesteryears*," 404.
228. Engelbrecht, 1157.
229. "Independent Hose Co. 175th," 29.
230. Deater and Deater, *History of Junior Fire Company*, 20.
231. FNP Archives, "The Emmitsburg Fire, Frederick Firemen Cover Themselves with Glory," *News*, March 21, 1885.
232. Ibid.
233. Ibid., "Creagerstown Fire Under Control after a Big Part Is Laid in Ruins," *Daily News*, June 2, 1914.
234. Ibid., "Sesqui-Celebration Held Despite Rain," *Daily News*, July 6, 1926.
235. Ibid., "Stirred by Disaster," *Daily News*, February 8, 1904.
236. Gary E. Frederick, *Fire Department Operations: A Technical Report*, "The Great Baltimore Fire," 101–3 (Lutherville: Fire Museum of Maryland, n.d.).
237. "Independent Hose Co. 175th," 29.
238. Ibid., 32.
239. Ibid., 30.
240. Ibid., 31.
241. FNP Archives, "Auto Engine Here, Independents Hose Company's New Machine Arrives," *Daily News*, September 2, 1908.
242. Ibid.
243. Ibid., "Plans for Independent Engine Test," *Daily News*, September 9, 1908.
244. Jewell and Jenkins, *Firefighting in Frederick*, 16.
245. FNP Archives, "Automobile Engine Hauled In," *Daily News*, October 6, 1908.
246. Ibid., "Ready for Calls," *Daily News*, December 1, 1908.
247. Ibid., "Spark Starts Fire, Big Blaze at J. Franklin Thomas Farm," *Daily News*, December 31, 1908.
248. Ibid., "County Property Menaced by Fire Saved by Hose Company," *Daily News*, April 29, 1924.

Chapter 10

249. Gordon and Gordon, "Swampers 150th," 24.
250. Deater and Deater, *History of Junior Fire Company*, 234.
251. Engelbrecht, 1143.
252. A copy of an article in the *Thurmont Clarion* of July 10, 1890, was provided to the author by Mayor John Kinnaird of Thurmont, who is also very involved in the preservation of history in Thurmont and a member of the Guardian Hose Company.
253. Brunswick Volunteer Company, www.bvfd5.org/content/history.
254. FNP Archives, "Local Firemen Fight Big Blaze in Burkittsville," *News*, July 13, 1925.
255. Ibid., "For Fire Protection, Things for Frederick to Do," *News*, July 2, 1904.
256. Information provided to author by the late John Maly, life member and historian of the Independent and member of the Frederick County Fire and Rescue Museum.
257. FNP Archives, "For Fire Protection, Recommendations Made by Board of Underwriters, Things for Frederick to Do," *Daily News*, July 2, 1904.
258. Deater and Deater, *History of Junior Fire Company*, 22.
259. Paul Ditzel, *Fire Alarm* (New Albany, IN: Fire Buff House Publishers Division of Conway Enterprises, Inc., 1990), 135.
260. Gordon and Gordon, "Swampers 150th," 4.
261. "Independent Hose Co. 175th," 3.
262. Jewell and Jenkins, *Firefighting in Frederick*, 60, 75, pictures of United 1919 Ahrens-Fox and a banquet invitation of the Citizens Truck Company.
263. FNP Archives, "Policy on Running Procedures Top Item for Fire Coordinator," *News*, April 22, 1980.
264. Ibid., "Want Use of Engine," *Daily News*, February 11, 1911.
265. Ibid., "Stubborn Blaze in Double Frame House," *Daily News*, September 6, 1915.
266. Ibid., "Bell and History Weekend Honors Citizens Truck Fire Company, Founded in 1926, Citizens Nears 50th Year," *Daily News*, May 18, 1972.
267. Ibid., "Fire Truck Meeting Tonight," *Daily News*, April 30, 1926.
268. Ibid., "Citizens Ladder Truck Co. Formed, New Fire Fighting Organization Has Charter Membership Over 300," *Daily News*, May 1, 1926.
269. Ibid., "Bell and History Weekend."
270. Ibid., "Support of Truck Co. Urged by Kiwanis Club," *Daily News*, May 26, 1926.

271. Ibid., editorial, "The Hook and Ladder Truck," *Daily News*, June 26, 1926.
272. Ibid., "City Petitioned to Buy Aerial Fire Equipment," *Daily News*, October 7, 1926.
273. Ibid., "$175,000 Bond Issue Permit Sought by City," *Daily News*, February 17, 1927.
274. Ibid., "Mack Co. Lowest Fire Truck Bidder, $1300 Between Low and High Price for Equipment," *Daily News*, April 22, 1927.
275. Ibid., "City Votes to Pay $15,000 for Truck Equipment, $1,500 to Be Made Up from Membership of New Company," *Daily News*, April 25, 1927.
276. Ibid., Local Mentions, "Resolution," *Daily News*, resolution adopted at a meeting of Independent Hose Company showing support to the Citizens Truck Company, September 1, 1927.
277. Artifacts at the Frederick County Fire and Rescue Museum: driver's hat and a badge with "A" as center, picture of 1952 American LaFrance with "A" in Maltese cross on door.
278. FNP Archives, "Official Test of Fire Truck Nov. 16," *Daily News*, November 14, 1927.
279. Ibid., "Exhibition Given at Truck Company Thursday," *Daily News*, November 25, 1927.
280. Ibid., "Citizens Truck Ready," *Daily News*, December 3, 1927.
281. Ibid., "Aldermen Take Up Fire Alarm System for City," *Daily News*, December 15, 1927.

Chapter 11

282. John Greenleaf Whittier, "Barbara Fritchie," www.poetryfoundation.org/poems/45483/barbara-frietchie. The opening stanza speaks of the "Clustered Spires of Frederick."
283. "Timeline of Trinity Chapel."
284. MD State Archives, 1805 Session Laws, vol. 607, p. 49, ch. 70, passed January 25, 1806.
285. Engelbrecht, 608.
286. Ibid., 464, 465.
287. Ibid., 466.
288. Ibid., 506.
289. Ibid., 606.
290. Ibid., 737.

291. "Timeline of Trinity Chapel," 5.
292. Ashbury, "*…And All Our Yesteryears,*" 148.
293. Ibid., 165.
294. Ibid., 125.
295. Gordon and Gordon, "Swampers 150th," 25.
296. "Timeline of Trinity Chapel," 5.
297. Gordon and Gordon, "Swampers 150th," 7.
298. FNP Archives, "New Fire Engine Tested by Juniors," *Daily News*, October 30, 1924.

Chapter 12

299. Ibid., "The Firemen's Association," *Daily News*, February 7, 1893.
300. Ibid., "The Independents, Arrangements in Progress for Their 75th Anniversary," *Daily News*, July 8, 1892.
301. Ibid., "Many Firemen Invited, Independent Anniversary Will Result in a State Organization," *Daily News*, July 21, 1892.
302. Ibid., "The Firemen's Association," *Daily News*, February 7, 1893.
303. Ibid., "Appointed Chief Marshal," *Daily News*, March 22, 1893.
304. Ibid., "More Convention Coming," *Daily News*, March 24, 1893.
305. Ibid., "Still They Come, Extensive Preparation for the Independent Anniversary in June," *Daily News*, January 6, 1893.
306. Ibid., "The Firemen Meet, Constitution of the State Association Adopted—Delegates Appointed," *Daily News*, March 28, 1893.
307. Ibid., "The Firemen's Demonstration," *Daily News*, May 29, 1893.
308. *Proceedings of the First Annual Convention of the Maryland State Firemen's Association, June 7th and 8th, 1893* (Frederick, MD: Press of Baughman Brothers, 1893).
309. Ibid.
310. MD Archives, 1894, 83, 84.
311. Convention Proceedings Book MSFA, 1893.
312. FNP Archives, "Farewell Firemen, the Last of the Monster Demonstration Departing for Their Homes," *Daily News*, June 9, 1893.
313. *75th Anniversary of the Independent Hose Co. No. 1 of Frederick City, Maryland, June 7th and 8th, 1893*, listing of parade winners, Frederick, Maryland, 1893.
314. "Maryland State Firemen's Association, Proceedings of the 126th Annual Convention and Conference," Ocean City, Maryland, June 16–22, 2018, 30.

Chapter 13

315. Engelbrecht, vol. 1, 29.
316. Ibid., 80.
317. Ibid., 137.
318. Ibid., 200.
319. Ibid., 209.
320. Ibid., 256.
321. Ibid., 349.
322. Ibid., 370.
323. Ibid., 542.
324. Ibid., 639.
325. Ibid., 641.
326. Frederick Mutual Insurance Company, www.frederickmutual.com/about-insurance-company.
327. Engelbrecht, 664.
328. Ibid., 683.
329. Ibid., 689.
330. Ibid., 731.
331. Ibid., 742.
332. Ibid., 753.
333. Ibid., 824.
334. Ibid., 708.
335. FNP Archives, "A Catastrophe, an Explosion Follows a Fire, Injured and Dead…," *Daily News*, May 29, 1888.
336. Ibid., "An Improvement," *News*, May 28, 1888.
337. Ibid., "Wind and Flame, a Serious Sunday Fire at Point of Rocks," *News*, March 24, 1890.
338. Gordon and Gordon, "Swampers 150th," 21.
339. FNP Archives, "Disastrous Fire at New Market, Episcopal Church and Two Dwellings Destroyed," *News*, May 2, 1902.
340. Ibid., "Destructive Fire Visits Mt. Airy, Fifteen Buildings, Including Two Hotels, Burned…," *News*, February 25, 1903.
341. Ibid., "Mt. Airy Fire Swept, Ten Buildings Destroyed—Loss $90,000—Business District Almost Wiped Out," *Citizen*, March 27, 1914.
342. Ibid., "Fire Sweeps Through Business Section of Mt. Airy, Nine Buildings Wrecked, Loss Estimated at $200,000," *News-Post*, June 5, 1925.
343. Marie Anne Erickson, *Frederick County Chronicles: The Crossroads of Maryland* (Charleston, SC: The History Press, 2012), 21.

344. FNP Archives, "Creagerstown Fire Under Control after a Big Part Is Laid in Ruins…," *News*, June 2, 1914.
345. "$30,000 Fire at Myersville, 5 Business Places Burned, Disastrous Blaze in Sub-Station of H&F Railway, about 1.30 Last Saturday Afternoon," *Valley Register* (Middletown, MD), copy of article in collection of author.
346. FNP Archives, "One Mountain Fire Under Control but That One Near Winchester Still Raging," *News*, November 25, 1922.
347. Ibid., "Costs $16.75 for Failing to Fight Fire in Mountain," *News*, May 15, 1922.

Chapter 14

348. Gordon and Gordon, "Swampers 150th," 13, 14.
349. Ibid., 13.
350. Items on display at the Frederick County Fire and Rescue Museum.
351. Gordon and Gordon, "Swampers 150th," 14.
352. Engelbrecht, 1115.
353. Ibid., 1147.
354. Gordon and Gordon, "Swampers 150th," 14.
355. Engelbrecht, 831.
356. FNP Archives, "Shame! Shame! Shame! Frederick Firemen Turn Hoodlums and Assault Each Other," *Daily News*, July 15, 1895.

Chapter 15

357. Myersville Volunteer Fire Company, www.myersvillevfc.org/history.
358. FNP Archives, "Fire Protection for New Midway, Enterprising Citizens to Organize to Install a Chemical Engine," *Frederick Post*, August 2, 1926.
359. Jewell, *Firefighting in Frederick County*, 36.
360. FNP Archives, "New Engine Tested, New Midway Equipment Delivered Last Week," *Frederick Post*, August 29, 1927.
361. Information on Jesse Renner's service on board of county commissioners, www.findagrave.com/memorial/66913866/jesse-isaac-renner.
362. Rocky Ridge, *50th Anniversary*, 3.
363. Carroll Manor Fire Company 50th Anniversary Book, *Carroll Manor Fire Company, 50 Years of Serving Our Community* (Adamstown, MD, 2003), 2, 3, 4.

364. FNP Archives, "New Market Residents Strong for Adequate Fire Protection," *Frederick Post*, March 21, 1918.
365. Ibid., "Engine for New Market," *Frederick Post*, April 10, 1918.
366. Ibid., "Engine for New Market," *Frederick Post*, April 26, 1918.
367. New Market District Volunteer Fire Company, www.nmvfc15.org/content/history.
368. Author interview with Donald L. Trimmer and documents Trimmer provided for review, October 21, 2020.
369. Jewell, *Firefighting in Frederick County*, 105.
370. Information on Lawrence A. Dorsey, www.findagrave.com/memorial/66913866/lawrence-dorsey.
371. Information from an ad placed by the Graceham Volunteer Fire Company in the 1938 Frederick County Volunteer Firemen's Association Annual Convention.
372. FNP Archives, "Firemen File Suit for Bonds," *News*, April 3, 1961.
373. Ibid., "Graceham Firefighters Mark Incorporation Anniversary," *Post*, April 29, 1965.
374. Ibid., "Firemen Turn Down Graceham, Also Set Up a Group to Set Up Central Alarm Fire Board at County Meeting," *Post*, October 18, 1963.
375. Ibid., "Firemen Seek Court Ruling, Graceham Company Wants Decree to Determine County Membership," *Post*, February 26, 1964.
376. Ibid., "New Graceham Fire Pump for Drought Area," *Post*, August 27, 1966.
377. Brunswick Volunteer Ambulance Company, bvar19.com/about.html.
378. Myersville Volunteer Fire Company, www.myersvillevfc.org/history.
379. Jewell, *Firefighting in Frederick County*, 77.
380. FNP Archives, "Fire Company Planned for Lewistown District," *News* April 28, 1967.
381. Jewell, *Firefighting in Frederick County*, 81
382. Urbana Volunteer Fire & Rescue, Inc., urbanavfd.com/about-us.
383. Walkersville Volunteer Rescue Company, www.wvrc24.com/content/history.
384. Thurmont Community Ambulance, www.thurmontambulance.org/history/history-menu.
385. Recollections of author as a Central Alarm dispatcher in early 1970s.
386. Vigilant Hose Company, www.vhc6.com/content/history.

ABOUT THE AUTHOR

Deputy Chief Clarence "Chip" Jewell retired in 2017 as director of the Frederick County Division of Volunteer Fire and Rescue Services. He serves as president and assistant chief of the Libertytown Volunteer Fire Department and is also a life member of the Junior Fire Company No. 2 in Frederick. He is a co-author of *Firefighting in Frederick* and author of *Firefighting in Frederick County*, both books in the Images of America series from Arcadia Publishing. He holds a bachelor's degree in fire science from the University of Maryland/ University College.

He is the president and a founder of the Frederick County Fire and Rescue Museum and Preservation Society, Inc. and was a founding member of the National Fire Heritage Center. He is vice chairman of the Historical and Archives Committee of the Maryland State Firemen's Association and a columnist for the MSFA publication *Volunteer Trumpet*. He is a monthly columnist for the *Frederick News-Post* and has written for *FIREHOUSE Magazine* and FIREHOUSE.com. He is a member of Heritage Frederick and has received the National Historic Preservation Award from the Daughters of the American Revolution.